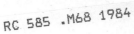

MOULD ALLERGY

MOULD ALLERGY

Edited by

YOUSEF AL-DOORY, Ph.D.

Associate Professor
Department of Pathology (Laboratory Medicine)
The George Washington University School of Medicine and Health Sciences
Washington, DC

and

JOANNE F. DOMSON, M.D.

Fellow in Allergy and Immunology
Children's Hospital National Medical Center
Washington, DC

Lea & Febiger   *Philadelphia*   *1984*

Lea & Febiger
600 South Washington Square
Philadelphia, Pa.   19106
U.S.A.

Library of Congress Cataloging in Publication Data
Main entry under title:

Mould allergy.

  Bibliography: p.
  Includes index.
  1. Allergy. 2. Fungi. I. Al-Doory, Yousef,
1924–   . II. Domson, Joanne F. [DNLM: 1. Hypersensitivity. 2. Fungi. WD 300 M926]
RC585.M68 1983   616.9′69   83-14951
ISBN 0-8121-0897-3

Cover illustration: Scanning electronmicrograph of conidia of *Cladosporium macrocarpum*. Courtesy of Dr. Garry T. Cole.

Printed in the United States of America

Print Number: 5   4   3   2   1

PREFACE

Not all textbooks on allergy discuss fungal allergy to any significant degree; even when it is mentioned, fungal allergy remains the least emphasized topic in most books on allergy. Over the past two decades most allergists have recognized a need for more information on fungal allergy because of increased awareness of the problem and the greater number of patients suffering from asthma and rhinitis due to fungi. Sources of information regarding fungal allergy and allergenic fungi, however, have been widely scattered, and the average practitioner in allergy usually does not have the time to search the literature for information on aspects of fungal allergy.

Most allergists are familiar with allergy to pollens, dust, animals, and food, and even to insect venoms. But seasonal variation in regard to fungi in the air or to those in homes, and their detection and monitoring, are generally new fields of study for the average allergist.

In this volume we have tried to bring together available data and information on fungal allergy so as to make it accessible to all with an interest in this field; in addition, we have supplied some basic information on fungi in general, and on those causing allergy in particular. We believed that the best way to reach our goal was to draft experts on different aspects of this topic to write the chapters. It was our hope that this would allow us to present pertinent data and current reports on the various facets of fungal allergy.

It is worth mentioning here that this book should have been called "Fungal Allergy" instead of "Mould Allergy," because the fungi include both moulds and yeasts, and because members of both groups can cause allergy. Most allergists, however, refer to fungal allergy as mould allergy. We used the term "mould" rather than "mold" because the latter is a "cast" and not a type of fungi.

Our thanks are extended to all who contributed manuscripts, to Deborah E. Anderson and Jno Randall for their valued efforts, especially in photography, to Joan Best and Paul Domson for their support, to Linda Jo Al-Doory for typing the manuscript and to the Division of Laboratory Medicine (Department of Pathology) of George Washington University School of Medicine for all the help and cooperation given to us while this text was being prepared. Our special gratitude is extended to Shirley Ann Ramsey for her valuable suggestions in reviewing the final edited version.

Washington, D.C.   Yousef Al-Doory, Ph.D.
                                                              Joanne F. Domson, M.D.

CONTRIBUTORS

Kjell Aas, M.D.   Pediatric Department
                              Rickshos Pitalet
                              Oslo, Norway

Yousef Al-Doory, Ph.D.   Department of Pathology
                              The George Washington University School of
                                  Medicine and Health Sciences
                              Washington, D.C.

Lars Aukrust, Ph.D.   Head, Biochemical Research Department
                              Nyegaard and Company
                              Oslo, Norway

Joseph J. Barboriak, M.D.   Department of Pharmacology
                              The Medical College of Wisconsin
                              Milwaukee, Wisconsin

Garry T. Cole, Ph.D.   Department of Botany
                              University of Texas
                              Austin, Texas

Joanne F. Domson, M.D.   Fellow in Allergy and Immunology
                              Children's Hospital National Medical Center
                              Washington, D.C.

Mervyn L. Elgart, M.D.   Chairman, Department of Dermatology
                              The George Washington University Medical
                                  Center
                              Washington, D.C.

Jordan N. Fink, M.D.   Allergy-Immunology Section
                              Department of Medicine
                              The Medical College of Wisconsin
                              Milwaukee, Wisconsin

Donald R. Hoffman, Ph.D.   Department of Pathology and Laboratory
                                  Medicine
                              School of Medicine
                              East Carolina University
                              Greenville, North Carolina

William A. Howard, M.D.   Clinical Professor of Child Health and
                          Development
                          The George Washington University School of
                          Medicine and Health Sciences
                          Director Emeritus
                          Department of Allergy and Immunology
                          Children's Hospital National Medical Center
                          Washington, D.C.

Peter P. Kozak, Jr., M.D.   Pediatric and Adult Allergy
                            Orange, California

Viswanath P. Kurup, Ph.D.   Allergy-Immunology Section
                            Department of Medicine
                            The Medical College of Wisconsin
                            Milwaukee, Wisconsin

Charles W. Mims, Ph.D.   Chairman, Department of Biology
                         Stephen F. Austin State University
                         Nacogdoches, Texas

Robert A. Samson   Centraalbureau Voor Schimmelcultures
                   AG Baarn, Netherland

Mark R. Sneller, Ph.D.   Aeroallergen Research
                         Tucson, Arizona

William R. Solomon, M.D.   Division of Allergy
                           Department of Medicine
                           The University of Michigan
                           Ann Arbor, Michigan

Gerald E. Wagner, Ph.D.   Department of Microbiology
                          The George Washington University School of
                          Medicine and Health Sciences
                          Washington, D.C.

CONTENTS

Chapter 1
INTRODUCTION

Yousef Al-Doory

Ancient sources reported allergic-type reactions in some patients who seemed to suffer from eating certain foods or from coming into contact with certain substances; at that time, these reactions were not systematically recognized as being allergic entities. A good example is Bostock, who in 1819, described his own suffering from a disease as "summer catarrh"; he renamed the disease "hay fever" nine years later. Bostock reasoned, although with some doubt, that this reaction was caused by exposure to new-mown hay. In 1873 Blackley demonstrated that Bostock's "hay fever" was actually caused by exposure to grass pollens (5).

Von Pirquet proposed the term "allergy" in 1906 to designate any altered reactivity in humans or animals due to the introduction of a foreign substance. A foreign substance was defined as an infectious agent or substance used for treatment of any such infection (6). Farr and Spector (3) introduced a new definition for allergy involving immunologic mechanisms in the pathophysiology. They defined allergy as "untoward physiologic events mediated by a variety of different immunologic reactions." Accordingly, three criteria would be needed for a definite diagnosis of an allergic state: 1) the identification of the allergen (antigen); 2) the establishment of a relationship between exposure to the allergen and occurrence of symptoms; and 3) the identification of the mechanism of the immunologic reaction involved in the illness (5).

Correlation of allergic responses to human infection provided a diagnostic leap in the clinical handling of pathogenesis from infectious diseases. For example, allergic reactions usually accompany infections with bacterial, viral, or fungal agents or with spirochetes. Allergies generally produce marked tissue reaction in the patient toward the infectious agent; thus, allergic reactions may influence determination of the cause of the pathology. In many instances allergic reaction provides the basis for invaluable diagnostic skin testing (6).

Antibodies form in the patient during a typical allergic reaction to an infectious agent. Antibodies are also found in other types of allergies, such as anaphylaxis, serum sickness, and the Arthus reaction. No antibodies have been detected, however, when the allergy is due to psychologic or physical causes, when no allergenic agents were introduced into the patient's body. Allergies due to cold, heat, sunlight, or even certain thoughts and daydreams do seem to occur but, without the introduction of an external substance into the patient's body, the body fails to produce antibodies. However, certain allergic and asthmatic symptoms may emerge in these cases. In addition, cases have been reported in which all clinical criteria pointed to an allergy, reasonably attributed to antigen-antibody reactions, but without the detection of antibodies. A good example of this type of reaction occurs when serum from patients with asthma-hay fever allergies is injected into normal human skin, producing local sensitization without antibodies in the blood stream of the recipient. This phenomenon can also be demonstrated in infective asthma, infective urticaria, and most drug allergies. It seems reasonable to assume that forms of antibodies are produced in these unusual cases, but new technical approaches are required to detect them.

Sherman and Kessler have suggested that the term "allergy" be restricted to phenomena of acquired, specific sensitization, characteristically consistent with antigen-antibody reactions, whether or not antibody can be demonstrated (6).

The relationship of fungi to asthma and hay fever was first noted in 1726, when Floyer reported that a patient suffered an asthma attack following a visit to a wine cellar (4). The first documented case of asthma due to a fungus was reported in 1924 (see Chapter 9).

THE FUNGI

The fungi are a group of eukaryotic, typically filamentous, spore-bearing organisms. Fungi originally were considered a group of plants with the entire body (yeasts or moulds) lacking any stems, leaves, or roots. They were classified along with the algae and the lichens within the class Thallophyta. Fungi differed from the other two groups, however, by their lack of chlorophyll. They had to depend on an outside source for nourishment, and they existed as saprobes or parasites. In recent years there has been a tendency to place the fungi in a separate kingdom, the Myceteae (see Chapter 2).

There are currently about 80,000 described species of fungi (2), both yeasts and moulds (Fig. 1–1), with predictably more species awaiting discovery. The yeasts typically contain single cells and multiply asexually by budding. The true yeasts, in addition to their asexual multiplication,

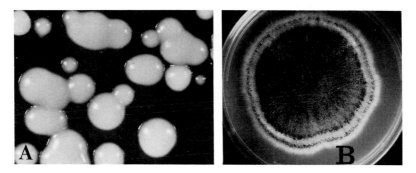

Fig. 1–1. *A.* Yeast colonies *(Candida parapsilosis). B.* A mould colony *(Aspergillus niger).* *(A* from Al-Doory, Y.: Laboratory Medical Mycology. Philadelphia, Lea & Febiger, 1980.)

multiply sexually by the formation of ascospores or basidiospores. Moulds are comprised of branching tubular hyphae. Hyphae may be divided by cross walls (septa) to form individual compartments or cells, or they may lack cross walls, in which case they resemble an open continuous tube with nuclei floating freely within the cytoplasm (Fig. 1–2). Such nonseptate hyphae are usually called coenocytic hyphae. Collectively, all hyphae of one fungal organism are called a mycelium. Certain pathogenic fungi are dimorphic: they exist in nature as moulds (spores) and convert to either yeasts or spherules after they enter and infect the body of an animal or human. Perhaps the best examples of this type of fungi are *Blastomyces dermatitidis* and *Coccidioides immitis* (Fig. 1–3). Furthermore some yeasts (referred to as yeastlike in certain textbooks), under special environmental conditions, including tissue invasion, may form

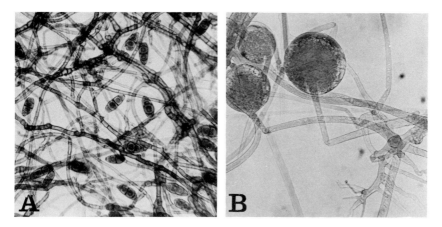

Fig. 1–2. *A.* Wet mount showing septate hyphae and conidia *(Curvularia* sp.). *B.* Wet mount showing nonseptate hyphae and sporangia *(Rhizopus* sp.).

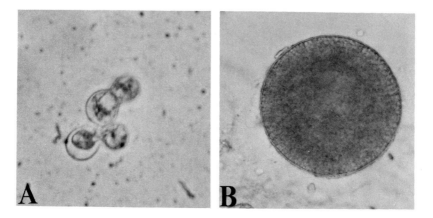

Fig. 1–3. *A.* Yeast phase of *Blastomyces dermatitidis. B.* A spherule of *Coccidioides immitis.* (From Al-Doory, Y.: Laboratory Medical Mycology. Philadelphia, Lea & Febiger, 1980.)

what is called pseudohyphae as well as blastospores (blastoconidiospores), arthrospores (arthroconidiospores), and chlamydospores, depending on the genus and species of the yeast (Fig. 1–4).

Fungi are important in human and environmental processes, functioning in ways that maintain ecologic purposes or enrich human existence or, on the other hand, destroy human well-being or life. Fungi are ubiquitous in the environment in large numbers, and are the organisms responsible for much of the disintegration of organic matter. They destroy food, fabrics, leather, and most consumer goods manufactured from raw materials. They are the causative agents of the majority of plant diseases, as well as human and animal diseases (1). Toxic and poisonous fungi are other examples of harmful ones.

In contrast, however, fungi are beneficial for agriculture, where as saprobes they are responsible, along with bacteria, for recycling impor-

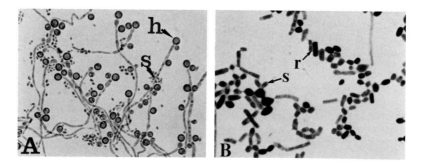

Fig. 1–4. *A.* Cornmeal culture of *Candida albicans* showing pseudohyphae, blastospores (s), and chlamydospores (h). *B.* Gram-stained smear of *Trichosporon* sp. showing blastospores (s) and arthrospores (r). (From Al-Doory, Y.: Laboratory Medical Mycology. Philadelphia, Lea & Febiger, 1980.)

tant chemical elements. Fungi are used as food, as producers of anti-biotics, as fermenters, and as sources of drugs, as well as in many aspects of industry. The role of fungi in allergy is well documented throughout this text.

Those fungi most responsible for causing allergy include species belonging to *Alternaria, Cladosporium (Hormodendrum), Aspergillus, Helminthosporium, Fusarium, Phoma, Epicoccum, Pencillium, Rhizopus, Mucor, Aureobasidium pullulans, Nigrospora, Scopulariopsis,* and the spores of rusts and smuts (Figs. 1–5 to 1–7). Other fungi have occasionally been reported as being allergenic.

Another group of microorganisms important to the field of allergy are the Actinomycetes. These organisms simulate fungi in branching fila-

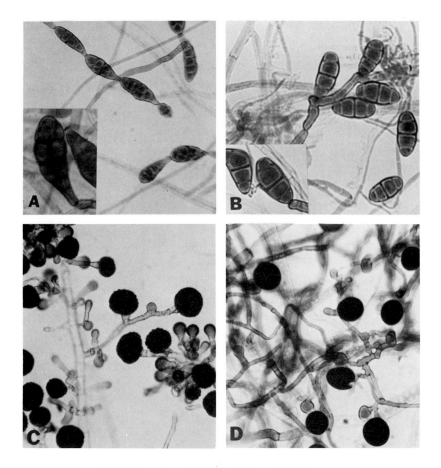

Fig. 1–5. Wet mounts of *Alternaria* sp. *(A), Curvularia* sp. *(B), Epicoccum* sp. *(C), Nigrospora* sp. *(D).* (From Al-Doory, Y.: Laboratory Medical Mycology. Philadelphia, Lea & Febiger, 1980.)

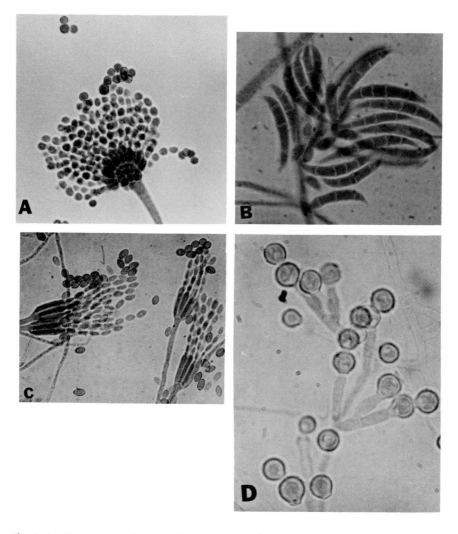

Fig. 1–6.   Wet mounts of *Aspergillus* sp. *(A)*, *Fusarium* sp. *(B)*, *Penicillium* sp. *(C)*, *Scopulariopsis* sp. *(D)*. (B–D from Al-Doory, Y.: Laboratory Medical Mycology. Philadelphia, Lea & Febiger, 1980.)

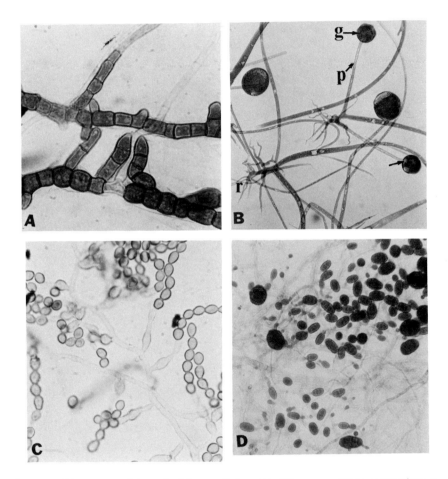

Fig. 1–7. Wet mounts of *Aureobasidium pullulans (A)*, *Rhizopus* sp. *(B)* (g, sporangium; p, sporangiophore; r, rhizoids; arrow, columella), *Cladosporium* sp. *(C)*, *Stemphylium* sp. *(D)*. (From Al-Doory, Y.: Laboratory Medical Mycology, Philadelphia, Lea & Febiger, 1980.)

ments (Fig. 1–8) and in the types of diseases they produce, but they are related more to bacteria than to fungi in size, physiologic reactions, and biochemical components. Among these are species that belong to the genera *Actinomyces, Nocardia,* and *Streptomyces.* The Actinomycetes of concern to allergists are those within the group of thermophilic agents. A species of *Thermoactinomyces* is an example of this group. Medical mycology laboratories study these microorganisms as they relate to human disease, and many medical mycology textbooks still include them in their discussion along with the other fungi. It is generally accepted at present, however, that Actinomycetes are filamentous bacteria rather than fungi.

Fungi reproduce by fragmentation, fission, and spore production. Except for the Deuteromycetes (fungi imperfecti), all fungi reproduce sexually as well as asexually (not necessarily simultaneously). Classification of fungi is mainly dependent on the spores' characteristics, with certain emphasis on sexual spores. This is apparent in the naming of certain classes of fungi: Basidiomycetes (Basidiomycota) produce basidiospores; Ascomycetes (Ascomycota) produce ascospores.

Asexual reproduction in yeasts, as previously mentioned, is characterized mainly by the formation of buds (with or without the formation

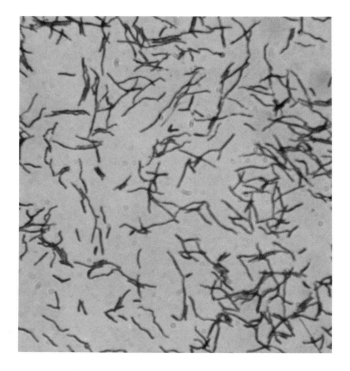

Fig. 1–8.   Gram-stained smear of *Actinomyces* sp. (From Al-Doory, Y.: Laboratory Medical Mycology. Philadelphia, Lea & Febiger, 1980.)

of blastospores, arthrospores, or chlamydospores) (Figs. 1–4 and 1–9). *Geotrichum* sp., as well as a few other moulds, also produce arthrospores (Fig. 1–10).

Most of the spores (conidia) formed by the imperfect fungi (Deuteromycetes) vary in shape, size, texture, color, number of cells, thickness of the cell wall, and method by which they are attached to each other and to their conidiophores (Figs. 1–5 to 1–7, 1–13).

Fungi can produce a large number of spores from a microscopic amount of fungal growth. *Aspergillus, Penicillium, Rhizopus, Mucor, Fusarium,* and *Gliocladium* are good examples (Fig. 1–12). The predominant characteristic of these fungi for producing a large number of spores is what makes them so important for study and monitoring: that is, they are present in large amounts in both the indoor and outdoor environment at all times. This is an important factor in the continued production of allergy in susceptible patients.

A microbiologist can usually identify most of the common fungi with ease. Speciation of fungi, as well as identification of uncommon ones, is more difficult, though, and should be handled by expert mycologists. It is important to remember that fungal colony characteristics, and even

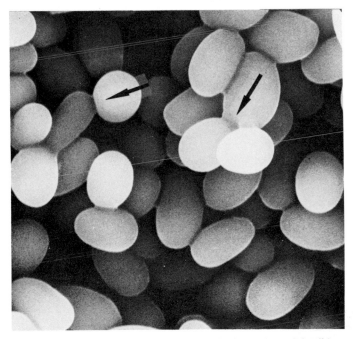

Fig. 1–9. Scanning electron micrograph of surface of colony of *Candida albicans,* showing details of bud formation *(arrows).*

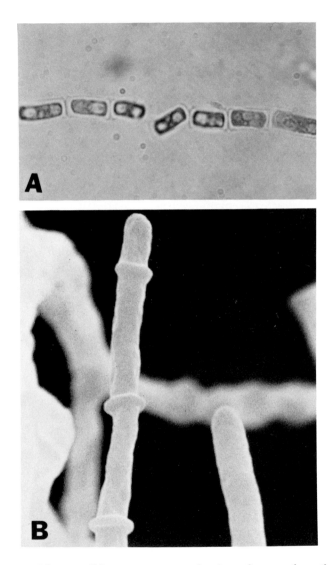

Fig. 1–10. *Geotrichum candidum. A.* Wet mount showing arthrospore formation. *B.* Scanning electron micrograph showing details of a hyphal element. (From Al-Doory, Y.: Laboratory Medical Mycology. Philadelphia, Lea & Febiger, 1980.)

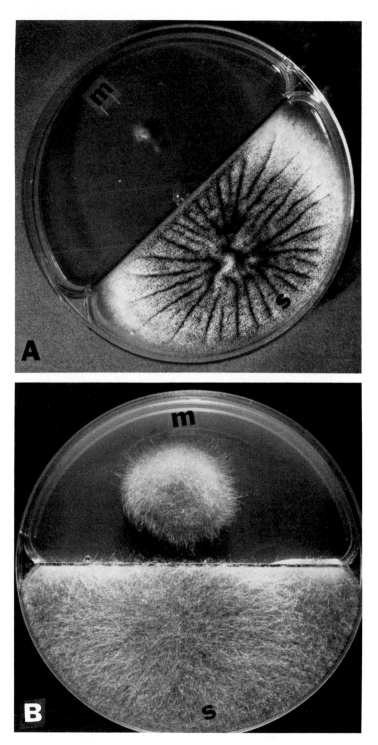

Fig. 1–11.   Effect of medium components on fungal growth. *A.* Inhibition of growth of *Aspergillus niger. B.* Reduction in growth rate of *Mucor* sp. (s, Sabouraud's dextrose agar; m, mycosel). (From Al-Doory, Y.: Laboratory Medical Mycology. Philadelphia, Lea & Febiger, 1980.)

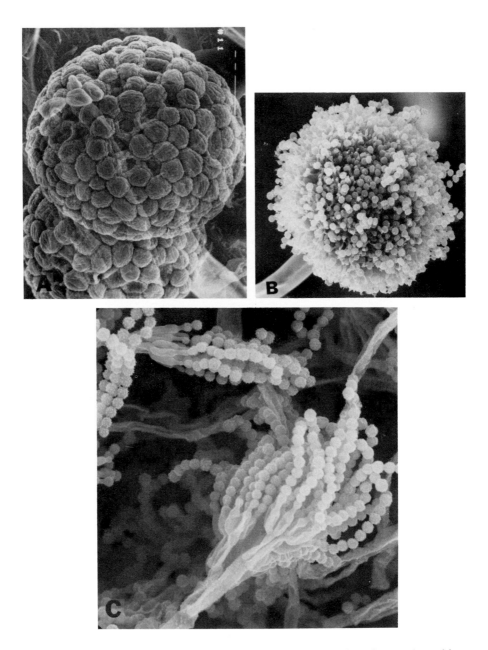

Fig. 1–12. Scanning electron micrographs showing the large number of spores formed by *Rhizopus* sp. *(A)*, *Aspergillus niger (B)*, *Penicillium* sp. *(C)*. *(B, C* from Al-Doory, Y.: Laboratory Medical Mycology. Philadelphia, Lea & Febiger, 1980.)

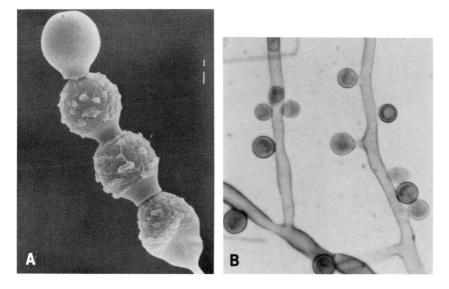

Fig. 1–13. Conidia formation by moulds. *A.* Scanning electron micrograph showing details of a chain of conidia of *Scopulariopsis* sp. *B.* Wet mount of *Monosporium apiospermum.* *(B* from Al-Doory, Y.: Laboratory Medical Mycology. Philadelphia, Lea & Febiger, 1980.)

microscopic characteristics, vary according to the medium on which the fungus is grown, the temperature of incubation, and the strain and variety of the fungal species (Fig. 1–11).

REFERENCES

1. Alexopoulos, C.J., and Mims, C.W.: Introductory Mycology. 3rd Ed. New York, John Wiley and Sons, 1979.
2. Bold, H.C., Alexopoulos, C.J., and Delevoryas, T.: Morphology of Plants and Fungi. New York, Harper and Row, 1973.
3. Farr, R.S., and Spector, S.L.: What is asthma? *In* The Asthmatic Patient in Trouble. Edited by T. Greenwich, Petty. C.P.C. Communications, 1975.
4. Floyer, J.: Violent Asthma After Visiting a Wine Cellar. A Treatise on Asthma. 3rd Ed. London, 1726.
5. Middleton, E., Jr., Reed, C.E., and Ellis, E.F.: Allergy. Principles and Practice. St. Louis, C.V. Mosby, 1978.
6. Sherman, W.B., and Kessler, W.R.: Allergy in Pediatric Practice. St. Louis, C.V. Mosby, 1957.

Chapter 2
CLASSIFICATION OF FUNGI

Charles W. Mims

The fungi are a very large and diverse group of organisms found in virtually every ecologic niche. Included here are forms commonly referred to as mushrooms, moulds, mildews, yeasts, rusts, smuts, bracket fungi, jelly fungi, cup fungi, puffballs, and stinkhorns. As a result of diversity the taxonomy of the fungi is complicated, and often there is disagreement even among specialists as to how various groups of fungi should be treated. As a result it is neither practical nor desirable to attempt to consider the taxonomy of the fungi in any great detail in a book of this nature. This chapter, therefore, will simply introduce the nonmycologist to the major groups of fungi and consider some problems associated with the classification of these interesting and important organisms.

CHARACTERISTICS OF FUNGI

Perhaps it is appropriate to begin this discussion by briefly considering the characteristics of the organisms in question. It should first be emphasized that all fungi lack chlorophyll and cannot manufacture their own food. Most fungi are saprobes, and as such are important as decomposers in ecosystems. The ability of some fungi to decay food, fabric, and wood products makes them economically important to humans. There are also many parasitic fungi that attack either plants or animals, including humans, and cause various degrees of injury to their hosts. Still others form mutualistic relationships with other living things, as are found in lichens and in mycorrhizae.

In addition to being heterotrophs, fungi are eukaryotic organisms that may reproduce by various means including fragmentation, fission, budding, and spore production. Both sexual and asexual spores are produced by most fungi, although not necessarily simultaneously. The basic structural unit in most fungi is the hypha. A hypha is a tiny, typically branched, tubular filament possessing a definite cell wall composed of chitin and other complex carbohydrates. Hyphae may be divided by cross walls, called septa, into individual compartments or cells. Such septa are, however, rarely solid. Typically one or more pores are present in each septum so that there is cytoplasmic continuity between adjacent hyphal cells. Collectively all the hyphae of a fungus comprise the mycelium, or body of the organism. It should be noted here that not all fungi produce hyphae. Some exist exclusively as single cells or as single cells with rhizoids. Others are dimorphic and, depending on environmental conditions, grow either in a single-celled so-called yeast phase or in a mycelial phase.

It should be emphasized that dimorphism is a characteristic or ability that is merely one example of a phenomenon in fungi known as pleomorphism. In simple terms, pleomorphism refers to the ability of an organism with a single genotype to produce two or more phenotypes. Although this is certainly not unique to fungi it does complicate their study, because in fungi the different phenotypes may be produced sequentially, simultaneously, or totally separate from one another. It is often impossible to predict what will occur in a particular species, and one form or spore-producing phase may or may not metamorphose into another. Pleomorphism is particularly troublesome as far as classification and nomenclature of fungi are concerned, because mycologists and other researchers must be able to identify a fungus in whatever form it is encountered. (For a discussion of some of these problems you may wish to consult Weresub and Pirozynski (21). The topic of pleomorphism has also been discussed recently by Carmichael (5).)

As a result of the pleomorphic nature of fungi, the terminology used to describe the various forms or spore-producing states has become confused. Attempts have recently been made to standardize much of the terminology, and the proposals of Hennebert and Weresub (8) will be followed here. The whole fungus in all its facets, forms, and potentialities, latent or expressed, will be referred to as the "holomorph." The sexual or perfect state will be termed the "teleomorph" while the term "anamorph" will be used to describe the asexual or imperfect state. More will be said about these terms subsequently.

In reference to the reproductive capacities of fungi noted above, it is important here to discuss briefly the nature of fungal spores. Simply defined, a spore is a microscopic reproductive unit. Essentially it is a fragment of an organism consisting of a minimum complement of essential cellular components, including at least a single nucleus surrounded by a cell wall (except in the case of flagellated spores, termed zoospores, which have no cell walls). The taxonomy of fungi is based to a large degree on spore characteristics. Important characteristics include not only spore size, shape, color, and surface ornamentation, but also spore ontogeny. Emphasis is, of course, placed on sexual spores, and many of the major classes of fungi are actually named on the basis of the type of sexual spore produced: basidiospores, class Basidiomycetes; ascospores, class Ascomycetes; zygospores, class Zygomycetes; and oospores, class Oomycetes. Fungal spores will be discussed further in Chapter 5, but at this point suffice it to say that the importance of spores in the taxonomy of fungi cannot be overemphasized.

TAXONOMY AND NOMENCLATURE

Taxonomy may be defined as the science of classification. As stated by Alexopoulos and Mims, the science

> has a dual purpose; first, to name organisms according to some internationally accepted system so that, with the least possible amount of confusion, mycologists may communicate to each other their findings concerning a certain fungus; second to indicate our current concept of the relationship of fungi to each other and to other living organisms (3).

The naming of species and the various other taxonomic categories comprise what is known as nomenclature. In naming fungi mycologists have chosen to follow the guidelines of the International Code of Botanical Nomenclature (ICBN). Some aspects of these guidelines are considered below. (For additional information regarding fungal systematics, taxonomy, and nomenclature, a good place to begin is Chapter 2 of Talbot's (19) Principles of Fungal Taxonomy.)

As has been the case with other living creatures, humans have attempted to give a scientific name to each type of fungus discovered. The scientific name of an organism is a binominal, the first word of which is a noun designating the genus to which the organism belongs. The second word, or specific epithet, is generally an adjective and denotes the species. Both names are Latin and should be underlined when written and italicized when printed. The genus is always capitalized and, although examples of the contrary may be found in older literatures, the modern tendency is not to capitalize the specific epithet. Binominals are usually derived from Greek or Latin roots and are often descriptive. Because mycologists adhere to the guidelines of the ICBN, an assignment of a name to a hitherto undescribed species requires publication of a Latin description. The formal description must be based on a so-called "type specimen" that should be deposited in a collection for future reference. The type is the original specimen on which the name is based. If designated by the author, it is the holotype; if chosen by a later author, it is the leucotype.

Following valid publication of a Latin description a formal citation will carry the name of the author, often in an abbreviated form, at the end of the specific epithet. The names and abbreviations of many authors of fungal taxa are available in Ainsworth (2). The date of publication of the description may also be added. Some binomials may be followed by two names, the first of which is in parentheses. The name in parentheses refers to the individual who first described the organism but who gave it a name different from that currently recognized. The second name is that of the person responsible for renaming the organism or for transferring it to a new or different genus. For example, let us consider the aquatic fungus *Aplanes treleaseanus* (Humphrey) Coker. This fungus was first described by Humphrey, who placed it in the genus *Saprolegnia*. Coker later determined that this fungus should be placed in the genus *Aplanes* and consequently changed the name. Name changes are, of course, quite troublesome, particularly to amateur mycologists and to biologists who work with fungi but who are not mycologists. Annoying as they may be, however, name changes are simply something we have had to learn to live with. See Miller and Farr (15) for a more detailed consideration of name changes in fungi.

In a number of instances different authors have unknowingly described the same fungus and have given it different names. Although each name may have been published according to ICBN guidelines, these specifications also note that a particular species may have only one scientific name. In a case in which synonyms exist the correct name for the species is the earliest that is in accordance with the rules. According to ICBN guidelines, names given to organisms prior to those given in Linnaeus' Species Plantarum in 1753 are not taken into account.

At this point note that, in attempting to apply the ICBN guidelines to fungi, certain problems have arisen because of the pleomorphic nature of fungi. Some of these have been considered in detail by Weresub and Pirozynski (21), who approached the topic from a historical standpoint. As noted previously, however, the basic problem is that many fungi produce separate sexual and asexual states that are often extremely difficult to link to one another. As a result special rules have been established for naming these fungi. Such is the case for the nonlichenized members of the classes Ascomycetes and Basidiomycetes. A brief consideration of these special nomenclatural rules is relevant here, because they apply to many fungi considered later in chapters of this book.

Under normal botanic rules for naming organisms, a botanic taxon is defined as an individual plant or organism. An individual organism is one in all its facets, forms, and potentialities, latent or expressed or, in the terminology of Hennebert and Weresub (8), the holomorph. Botanic names therefore apply to the holomorph even though the type specimen of the name may not be holomorphic material. It is in regard to the nature of the type specimen that special rules have been formulated for nonlichenized Ascomycetes and Basidiomycetes.

For most ascomycetous and basidiomycetous fungi to qualify for botanic names, the type specimen must at least be teleomorphic material or, in other words, must show evidence of sexuality. If the organism had first been described according to anamorphic or asexual material, the name given to the organism is a so-called form-name, or anatomic name (7), and applies only to that form or state of the fungus. It cannot be used to refer to the entire organism if additional forms or states are ever discovered. In other words, if a fungus initially described from its asexual state is later found to produce either ascospores or basidiospores (sexual spores), then the fungus must be renamed on the basis of the sexual state. The new name is the botanic name and applies to all forms or states of the fungus, including the asexual state. Because many Ascomycetes and a few Basidiomycetes were first described on the basis of their anamorphs, a situation exists in which a single fungus is known by two separate names, an anatomic or form-name and a botanic name. The botanic name has priority over the form-name although it is often useful to indicate both names in a citation. For example, the human pathogen *Cryptococcus neoformans* is a form-taxon named according to its anamorphic state. This fungus has, however, since been shown to have a basidiomycetous sexual state and has been renamed as *Filobasidiella neoformans* by Kwon-Chung (12). The latter name is the botanical name and therefore has priority over the former, although it is still useful to refer to the fungus as *Filobasidiella neoformans* (= *Cryptococcus neoformans*). In describing the asexual state of this fungus another option is simply to refer to the *Cryptococcus* state of *Filobasidiella neoformans*.

The anatomic system of classification using form-names also applies to the so-called "fungi imperfecti," or those fungi known only from their asexual states. In this discussion these fungi will be placed in the form-class Deuteromycetes. Generally these fungi are considered as asexual states of Ascomycetes or, in a few cases, as Basidiomycetes whose sexual states either do not exist or have not yet been discovered. Under botanic rules these fungi do not qualify for botanic names and are all assigned to form-taxa. Many fungi discussed elsewhere in this text actually belong in this category, and many common genera such as *Aspergillus, Penicillium*, and *Fusarium* are actually form-genera.

MAJOR TAXONOMIC CATEGORIES

Although the fungi have traditionally been placed in the plant kingdom by most researchers, the recent tendency has been to separate them into a kingdom of their own (20). This approach is followed here, in which the fungi are placed into the kingdom Myceteae. The kingdom is further divided into standard taxonomic categories, as follows:

Division
　Class
　　Order
　　　Family
　　　　Genus
　　　　　Species

Each of these categories may be subdivided as necessary into subdivisions, subclasses, suborders, and so forth. In accordance with the recommendations of the ICBN the names of divisions of fungi should end in -mycota, subdivisions in -mycotina, classes in -mycetes, subclasses in -mycetidae, orders in -ales, and families in -aceae. Genus and species names have no standard endings.

As previously noted there has been considerable disagreement regarding classification schemes for the fungi. Most mycologists have traditionally embraced whatever scheme is most suitable for them and, for the most part, have managed to communicate with each other successfully. In regard to the diverse opinions regarding taxonomic categories, perhaps the following quotation from Alexopoulos and Mims (3) is appropriate here:

> There is a tendency for the beginning student to regard these various taxonomic categories as concrete and stable, and more or less sacred. Such an attitude will lead to disappointment with the first attempt to identify an unknown organism. You should understand, above all, that living organisms are constantly evolving and that any attempts to pigeonhole them into a system of classification are nothing more than the attempts of biologists to organize their knowledge of the moment, and are strictly artificial. Even when our knowledge of fungi becomes much greater than it is at present, any attempt to draw hard and fast lines between taxonomic categories will be futile because the categories themselves are only human concepts and intermediate forms are bound to exist and to rise by hybridization and mutation.

The general taxonomic outline used in this chapter is shown in Table 2–1. It is probably no better than those of other systems, but hopefully will serve to introduce the reader to the various major groups of fungi. Other taxonomic schemes that you may wish to consider include those of Ainsworth (2) and Talbot (19).

As is evident in Table 2–1, Alexopoulos and Mims (3) have divided the kingdom Myceteae into three divisions: Gymnomycota, Mastigomycota, and Amastigomycota. The Gymnomycota contains those organisms producing somatic structures devoid of cell walls and exhibiting phagotrophic nutrition. Many biologists consider these organisms to be Protista but traditionally they have been studied primarily by mycologists, hence the tendency to include them with the fungi. The division contains three classes, the largest and probably best known of which is the Myxomycetes. This class contains those organisms commonly referred to as the "true plasmodial slime moulds." The somatic phase in

TABLE 2–1.   Divisions, Subdivisions, and Classes of the Kingdom Myceteae Recognized by Alexopoulos and Mims (3)

Kingdom Myceteae
　Division I. Gymnomycota
　　Subdivision 1. Acrasiogymnomycotina
　　　Class 1. Acrasiomycetes
　　Subdivision 2. Plasmodiogymnomycotina
　　　Class 1. Protosteliomycetes
　　　Class 2. Myxomycetes
　Division II. Mastigomycota
　　Subdivision 1. Haplomastigomycotina
　　　Class 1. Chytridiomycetes
　　　Class 2. Hyphochytridiomycetes
　　　Class 3. Plasmodiophoromycetes
　　Subdivision 2. Diplomastigomycotina
　　　Class 1. Oomycetes
　Division III. Amastigomycota
　　Subdivision 1. Zygomycotina
　　　Class 1. Zygomycetes
　　　Class 2. Trichomycetes
　　Subdivision 2. Ascomycotina
　　　Class 1. Ascomycetes
　　Subdivision 3. Basidiomycotina
　　　Class 1. Basidiomycetes
　　Subdivision 4. Deuteromycotina*
　　　Class 1. Deuteromycetes*

*Form-categories.

a myxomycete is a multinucleate acellular mass of protoplasm, a plasmodium, that moves and feeds in an amoeboid fashion. Eventually the plasmodium is converted to one or more fruiting bodies or sporophores that contain masses of powdery spores, each of which possesses a definite cell wall. The sporophores of these organisms are relatively easy to find in nature, and the spores of Myxomycetes sometimes show up in air samples. The plasmodia, on the other hand, usually go unnoticed, except to the eye of the experienced observer. *Physarum polycephalum* is the best known organism belonging to this group and has been studied extensively by developmental and cellular biologists. The authoritative work on the taxonomy of the Myxomycetes is that of Martin and Alexopoulos (14).

The other two classes belonging to the division Gymnomycota are the Protosteliomycetes and the Acrasiomycetes. The Protosteliomycetes are minute creatures that were only first described in 1960 by Olive and Stoianovitch (17). All the known species are invisible to the naked eye and are known only from laboratory culture. Most of what is known about this group can be found in Olive's The Mycetozoans (16). The Acrasiomycetes are also minute organisms, although they are more well known than the Protosteliomycetes. Like some of the Myxomycetes, certain Acrasiomycetes have also become popular research tools of the experimental biologist. One of the most popular species for experimental studies is *Dictyostelium discoideum;* many of the attributes that make it a popular developmental system are discussed by Loomis (13). Commonly known as "cellular slime moulds," the Acrasiomycetes have a somatic phase consisting of uninucleate naked cells that aggregate to form a pseudoplasmodium, which eventually produces a sorocarp bearing spores at its tip. During both aggregation and sporulation the naked cells maintain their individuality yet cooperate as a functional unit.

The division Mastigomycota contains fungi capable of producing flagellated cells at some point in their life cycle, and typically exhibiting absorptive nutrition. Some forms are unicellular though most produce an extensive mycelium consisting of aseptate hyphae. The division contains the classes Plasmodiophoromycetes, Hyphochytridiomycetes, Chytridiomycetes, and Oomycetes. The Hyphochytridiomycetes are aquatic fungi that produce anteriorly uniflagellate zoospores having a so-called "tinsel" flagellum. They parasitize either algae or fungi or grow saprobically on organic debris in the water. The class is a small one, and need not concern us here. Likewise, the Plasmodiophoromycetes and the Chytridiomycetes will probably be of little interest to you. The former group consists of obligate endoparasites of vascular plants, algae, and fungi. Only two species are of any economic importance, one attacking cabbage and the other potatoes. Members of the Chytridiomycetes are much more numerous than those of either the Hyphochytridiomycetes or the Plasmodiophoromycetes, and are characterized by the fact that they all produce motile cells equipped with a single posterior whiplash flagellum. Most of these fungi exist in aquatic habitats, although many are also found in the soil; most are saprobic, although a few parasitize various plants. One genus, *Coelomomyces,* contains species that attack mosquito larvae. Perhaps the best known chytrid genus is *Allomyces.* Numerous members of the genus have been studied extensively in the laboratory, and much is known of their cytology, physiology, genetics, and morphogenesis.

The largest class belonging to the division Mastigomycota is the class Oomycetes. These fungi are commonly referred to as "water moulds," although not all forms are aquatic. Most reproduce asexually by zoospores bearing two flagella, one tinsel and one whiplash. Some of the terrestrial plant-pathogenic Oomycetes produce propagules commonly termed conidia (naked nonmotile asexual spores), which are actually sporangia capable of producing zoospores if conditions are correct. Sexual reproduction in Oomycetes involves the interaction of a male sex organ, the antheridium, and a female sex organ, the oogonium. Following fertilization a thick-walled spore, termed an oospore, develops. Taxonomically the Oomycetes are often separated from other fungal groups because their thallus is diploid and their cell walls contain some cellulose. In most other fungal groups the thallus is haploid and cellulose is absent from the cell wall. A number of important plant pathogenic fungi, including various species of *Pythium, Phytophthora, Peronospora,* and *Plasmopara,* are Oomycetes, and as a result the class is of considerable economic importance.

The Amastigomycota is by far the largest of the three divisions recognized by Alexopoulos and Mims (3), and includes those fungi that exhibit absorptive nutrition and lack the ability to form flagellated cells. Most forms included in the class produce extensive well-developed mycelia, consisting either of septate or aseptate hyphae, although some single-celled organisms are also placed here. Included in the division are the classes Zygomycetes, Trichomycetes, Ascomycetes, and Basidiomycetes, as well as the form-class Deuteromycetes. All but the Trichomycetes, a group of fungi obligately associated with living arthropods, contain organisms of interest discussed in subsequent chapters. Each of the remaining four classes is briefly considered below.

The Zygomycetes are commonly referred to as the "bread moulds," although some soil-inhibiting mycorrhizal forms also belong here, as do a group of organisms that either prey on or parasitize amoebae, rhizopods, and nematodes. Most Zygomycetes produce well-developed hyphae that generally lack regularly occurring cross walls. Some of these fungi are dimorphic and can also exist in a yeast form if environmental conditions are satisfactory. In regard to nutrition most Zygomycetes are saprobes, although some parasitize other organisms, including humans. Examples of human pathogens include various species of *Rhizopus, Absidia, Mucor,* and *Basidiobolus.* A number of Zygomycetes are, on the other hand, beneficial to humans and have been used commercially in the production of various chemical compounds. Various species of *Rhizopus, Mucor,* and *Actinomucor* are also important in the production of fermented foods popular in the Orient.

Most researchers consider the Zygomycetes to be a well-defined natural group of fungi. Asexual reproduction in the group is typically by spores produced in a container called a sporangium. The asexual propagules of some Zygomycetes have been termed conidia, although recently the tendency has been to avoid using this term in reference to these structures. Sexual reproduction in the group is by zygospores. A zygospore is a thick-walled spore that is formed following the fusion of two equal or unequal gametangia. Interestingly, zygospores do not appear to be commonly produced, and a number of species of Zygomycetes have been described for which sexual reproduction is unknown. Unlike the situation described for Ascomycetes and Basidiomycetes, one of these organisms can be given, under ICBN guidelines, a botanic name based on a type specimen that is anamorphic material. For a detailed consideration of the classification of the Zygomycetes, refer to Benjamin (4).

By far the largest class of the division Amastigomycota is the class Ascomycetes. Included here are those yeasts that produce asci and ascospores, the plant-parasitic powdery mildews, the so-called cup fungi, the edible morels and truffles, and many other interesting and important fungi, including many that attack humans. All members of the class produce microscopic saclike structures called asci, in which sexually derived spores, ascospores, are produced. Although asci may arise directly from the fusion of simple cells, they generally arise after the interaction of a male reproductive structure, the antheridium, and a female structure, the ascogonium. In some forms the asci may be produced naked but they are usually produced in or on the surface of some sort of fruiting body, generally called an ascocarp. Although the ascocarps of most of these fungi are small and inconspicuous those of some genera, such as *Morchella* (the edible morel), are quite large.

Some Ascomycetes exist as single cells or produce only a scanty mycelium, but most of these fungi produce well-developed septate hyphae. The septa typically possess a simple central pore that allows cytoplasmic continuity between adjacent hyphal compartments. Some mycelial forms can also grow in a yeast phase, as in the case of certain human pathogens. Examples include *Ajellomyces dermatitidis (Blastomyces dermatitidis)*, which causes blastomycosis, and *Ajellomyces capsulatus (Histoplasma capsulatum)*, which causes histoplasmosis. Although the yeast forms reproduce asexually by either budding or fission, most Ascomycetes produce conidia. In fact, it is in this class that conidial formation reaches its zenith, as indicated by the tremendous diversity of conidial types produced by ascomycetous fungi. More will be said about conidia in reference to the form-class Deuteromycetes.

When discussing the Ascomycetes note that some of these organisms are more likely to be encountered in an asexual or conidial state than in a sexual or ascus-producing state. As a result an investigator sometimes faces the problem of trying to identify one of these fungi from its asexual state alone. Often this is difficult, and in many instances it is simply impossible because the conidia of many Ascomycetes resemble those of Deuteromycetes, or those fungi known only from their asexual states. In other words, unless the sexual state is available for examination, many Ascomycetes are morphologically indistinguishable from Deuteromycetes. This problem has, of course, been noted above, and the non-lichen-forming Ascomycetes are among those fungi for which special nomenclatural rules have been established. As you will recall, the valid scientific name (the botanic name) for one of these fungi must be based on a type specimen that is at least teleomorphic material. If the fungus has first been named on the basis of anamorphic material then the name given to the organism is a so-called form-name. The form-name applies only to the state or form of the fungus exhibited by the type specimen. Because many Ascomycetes were first described according to their asexual states it is not uncommon for one of these fungi to be known in the literature by two names, a form-name and a botanic name. The botanic name, of course, has priority over the form-name.

Kendrick and DiCosmo (10) have compiled an extensive list of teleomorph-anamorph connections in Ascomycetes that may be of interest. For example, all members of the ascomycetous genus *Nannizzia* have conidial stages resembling those of the form-genus *Microsporum*, while most members of the genus *Arthroderma* have *Trichophyton*-like conidia. Connections between teleomorphs and anamorphs are, however, not always as consistent as in these two examples. For instance, the various species of *Ceratocystis* key out to over ten different form-genera.

Before leaving the Ascomycetes note that there is little agreement among specialists as to how this class should be subdivided taxonomically. Basically, the problem is that there is no general consensus as to what constitutes the most reliable criteria of relationship. Some place major emphasis on ascus structure, while others argue that the totality of structures enclosed by the ascocarp wall should be considered. Still others attempt to consider both the teleomorph and anamorph or, in other words, the entire organism, or holomorph.

The class Basidiomycetes contains those fungi with which the layman is most familiar. Included here are forms such as mushrooms, puffballs, stinkhorns, bracket fungi, and jelly fungi, as well as the plant-pathogenic rusts and smuts. At some point in their life cycle all these fungi produce specialized structures, basidia, on the surface of which sexually derived basidiospores are produced. In most instances the basidia are borne on a macroscopic structure technically known as a basidiocarp. A mushroom with its cap, stalk, and gills is, for instance, a type of basidiocarp. The mycelium comprising the bulk of the organism producing the basidiocarp is usually not visible to the naked eye, because it permeates the substrate from which the fungus obtains its nutrients. Mycelia of most Basidiomycetes are extremely well developed, consisting of septate hyphae. Some Basidiomycetes possess a characteristic septal apparatus, referred to as the dolipore septum. Small structures known as clamp connections are also present on the hyphae of some Basidiomycetes.

Asexual reproduction in basidiomycetous fungi may be accomplished by various means, including the production of conidia. The asexual cycle does not, however, appear to play as significant a role in the life cycle of most of these fungi, as is the case in members of the classes Zygomycetes and Ascomycetes described above. Examples of Basidiomycetes of interest in this book include the rusts and smuts, whose spores are found routinely in surveys of airborne fungi. *Cryptococcus* and *Rhodotorula* are examples of form-genera that have basidiomycetous affinities.

The final taxonomic category to be discussed here is the form-class Deuteromycetes, or the "fungi imperfecti." As the designation "form-class" indicates, this is an artificial group consisting of those fungi known to reproduce only by asexual means. Although a few yeasts belong here, most of these fungi have well-developed septate hyphae and reproduce by conidia. These conidia may be produced directly on the hyphae, or within some sort of fructification or conidioma, to use the terminology of Kendrick and Nag Raj (11).

Taxonomically the Deuteromycetes hang like the proverbial albatross around the neck of the mycologist. As should be apparent by now, mycologists have essentially a dual system for naming and classifying fungi that belong to certain groups. The first is the botanic system for taxa of holomorphic fungi, while the second is the anatomic system for form-taxa of anamorphs or, in other words, deuteromycetous fungi. Enough has already been said concerning the problems of this dual system, but for some insight into how and why it evolved you are strongly encouraged to consult Weresub and Pirozynski (21) and Hennebert and Weresub (8). The history of conidial fungi has also been considered recently by Kendrick (9).

The taxonomy of the Deuteromycetes is based primarily on conidial characteristics. This is, of course, the so-called saccardoan system, which had its beginning in the nineteenth century. Basically, Saccardo (18) divided the Deuteromycetes into morphologic categories based on the shape, color, and septation of mature conidia. Descriptions of these categories are available in almost any introductory mycology text, and many terms used in this system have been more carefully defined recently by Kendrick and Nag Raj (11). Over the years many authors have expressed dissatisfaction with this system and have even proposed alternative approaches, but the saccardoan approach is still widely used today and currently is the best system for identifying a fungus according to its conidia. The major advantages of this system are its relative simplicity and its comprehensive nature.

As a group, the Deuteromycetes are of considerable interest and importance to those concerned with mould allergies. Included in the form-class are many organisms routinely found in surveys of airborne fungi. Examples include members of the form-genera *Alternaria, Penicillium, Aspergillus, Cladosporium, Helminosporium, Fusarium, Gliocladium, Stemphylium, Phoma, Scopulariopsis, Epicoccum, Trichoderma, Nigrospora, Rhodotorula,* and *Cryptococcus. Trichosporon, Pityrosporum, Microsporum,* and *Trichophyton* are also form-genera associated with either the hair or skin of humans. The human pathogens *Candida albicans* and *Coccidioides immitis* are also deuteromycetous fungi.

In closing it should again be noted that the conidial stages of many deuteromycetous fungi are often similar to those of Ascomycetes or, more rarely, of Basidiomycetes. As a result it is generally believed that these fungi represent Ascomycetes or Basidiomycetes whose sexual stages either have not been discovered or have been dropped from the life cycle during their evolution. It is, of course, also possible that some of these fungi may never have possessed a sexual state. For a detailed and comprehensive consideration of these fungi, refer to Biology of Conidial Fungi, edited by Cole and Kendrick (6).

REFERENCES

1. Ainsworth, G.C.: Ainsworth and Bisby's Dictionary of the Fungi. 6th Ed. Kew, Commonwealth Mycological Institute, 1971.
2. Ainsworth, G.C.: Introduction and keys to higher taxa. *In* The Fungi. Vol. IV B. Edited by G.C. Ainsworth, F.K. Sparrow, and A.S. Sussman. New York, Academic Press, 1973.
3. Alexopoulos, C.J., and Mims, C.W.: Introductory Mycology. 3rd Ed. New York, John Wiley and Sons, 1979.
4. Benjamin, R.K.: Zygomycetes and their spores. *In* The Whole Fungus. Vol. 2. Edited by B. Kendrick. Ottawa, National Museums of Canada, 1979.
5. Carmichael, J.W.: Pleomorphism. *In* Biology of Conidial Fungi. Vol. 1. Edited by G.T. Cole and B. Kendrick. New York, Academic Press, 1981.
6. Cole, G.T., and Kendrick, B.: Biology of Conidial Fungi. Vols. I and II. New York, Academic Press, 1981.
7. Hennebert, G.L.: Pleomorphism in fungi imperfecti. *In* Taxonomy of Fungi Imperfecti. Edited by B. Kendrick. Toronto, University of Toronto Press, 1971.
8. Hennebert, G.L., and Weresub, L.K.: Terms for states and forms of fungi, their names and types. Mycotaxon, **6**:207, 1977.
9. Kendrick, B.: The history of conidial fungi. *In* Biology of Conidial Fungi. Vol. 2. Edited by G.T. Cole and B. Kendrick. New York, Academic Press, 1981.
10. Kendrick, B., and DiCosmo, F.: Teleomorph-anamorph connections in ascomycetes. *In* The Whole Fungus. Vol. 1. Edited by B. Kendrick. Ottawa, National Museums of Canada, 1979.
11. Kendrick, B., and Nag Raj, T.R.: Morphological terms in fungi imperfecti. *In* The Whole Fungus. Vol. 1. Edited by B. Kendrick. Ottawa, National Museums of Canada, 1979.
12. Kwon-Chung, K.J.: A new genus, *Filobasidiella*, the perfect state of *Cryptococcus neoformans*. Mycologia, **67**:1197, 1975.
13. Loomis, W.F.: *Dictyostelium discoideum*. A Developmental System. New York, Academic Press, 1975.
14. Martin, G.W., and Alexopoulos, C.J.: The Myxomycetes. Iowa City, Univ. of Iowa Press, 1969.
15. Miller, O.K., and Farr, D.J.: An Index of the Common Fungi of North America (Synonomy and Common Names). J. Cramer, 1975.
16. Olive, L.S.: The Mycetozoans. New York, Academic Press, 1975.
17. Olive, L.S., and Stoianovitch, C.: Two new members of the Acrasiales. Bull. Torrey Bot. Club, **87**:1, 1960.
18. Saccardo, P.A.: Sylloge Fungorum Omnium Huzusque Cognitorum. Vol. 14. Pavia, published by the author, 1899.
19. Talbot, P.H.B.: Principles of Fungal Taxonomy. New York, St. Martin's Press, 1971.
20. Whittaker, R.H., and Margulis, L.: Protist classification and the kingdoms of organisms. Biosystems, **10**:3, 1978.
21. Weresub, L.K., and Pirozynski, K.A.: Pleomorphism of fungi as treated in the history of mycology and nomenclature. *In* The Whole Fungus. Vol. 1. Edited by B. Kendrick. Ottawa, National Museums of Canada, 1979.

Chapter 3
AIRBORNE FUNGI

Yousef Al-Doory

Fungi are found wherever there is organic matter that serves as a source of food supply. Fungi are omnivorous and ubiquitous as a group; also, some species are not limited in association with geography or food supply.

There are two principal reservoirs for fungus populations, water and soil. Sparrow (79) has described fungal species found most commonly in water in North America. From these species a few migrated into soil and formed specialized and rather small groups of fungal populations. The members (species) of the genus *Pythium* are good examples (55). The fungi of this aquatic group are known to prefer clean water with high levels of dissolved oxygen, but a few species have been found in polluted water.

As fungi became more complex and developed further morphologically, thus requiring additional nutrients, they moved toward land and their dependence on water was reduced. They retained the need for high relative humidity, however, which is required for fungal growth and for spore germination. These fungi either populate in specialized habitats or remain in the soil where they subsist on organic matter, such as leaf litters, compost and decaying herbage, carrion, fecal matter, or other available organic materials. Such fungi are usually known as soil fungi or as geofungi. Most of these fungi are discussed by Gilman (24).

Some land fungi became specialized in nutrient requirements, including such fungi as those that depend on dung or fecal matter of herbivorous or carnivorous animals (9,11). Furthermore, yeasts developed morphologically after their arrival on land; because of various conditions, however, some returned to a water habitat (69).

Other soil fungi were able to adapt to human habitats and activities and developed symbiotic relationships with humans (or animals), in which the fungi exist as part of the normal flora; some fungi, however, live as parasites in humans' (and other animals') bodies.

Examples of the ability of fungi to adapt to their surrounding environments within the soil are provided by the presence and role of fungi in mill slime (54), the decomposition of cotton and other celluloid fabrics (76), the production of toxic material in food (86), and their adaptation to certain nematodes and pests (17).

It would appear that whatever the fungal environment, soil or water, animal or plant, the fungi are able to survive, nourish, grow, germinate, and produce spores. The spores (when produced), whether formed in water, soil, on the surface of plants, or in debris, will be in contact with surrounding air, and therefore will become airborne. Climatic conditions are important in the distribution of airborne fungal spores throughout the atmosphere, as well as in the return of the spores to soil, water, or ground matter.

Accordingly, all atmospheric air, whether indoor or outdoor, on the surface of the ground or in high altitudes, contains certain varieties and amounts of fungal spores in addition to other atmospheric particles. The concentration of the fungal spores differs according to location, altitude, time of day, season of the year, condition of the surrounding area (e.g., farm, ocean, stable, downtown area), and climatic conditions (e.g., temperature, sunshine, humidity, rain, snow, wind, wind speed). Therefore fungal allergy sufferers are always exposed to fungal spores, and what differentiates exposure in one area from exposure in another is the quantity of spores in the air.

Airborne fungal spores may be of different types. Some may be pathogenic (e.g., *Coccidioides immitis*), some may be toxic (e.g., *Aspergillus flavus*), some may be harmless, but the most important ones to be considered in this chapter are those that are allergenic. Unfortunately, most of the allergenic fungi (if not all) are present in the atmospheric air throughout most fair weather months. Spore concentration may be reduced when there is a great deal of snow or rain, or in severely cold weather. The allergenic fungi are always present, however, and the primary question involves the significance of the number of spores in the air, which depends on the time of the day, season, weather, and surrounding environment.

It is safe to consider the soil as the main repository for all aerobic fungi. Fungal spores disperse from the soil into the air and then fall onto vegetation (causing disease); are airborne or otherwise carried into homes (causing, for instance, mouldy bathrooms and basements); are inhaled by humans and animals (causing toxic reactions, disease, allergy, or other fungal disorders), fall onto leather, wood, or food, for example (causing various mould damage), or fall back to the soil or onto other supportive materials and repeat the cycle. In any event, however, fungi must find a source of nourishment, whether parasitically or saprobically, for they cannot produce their own food. Similarly, it is important to remember that airborne fungal spores must be viable to produce disease or to grow and germinate in plants, animals, or humans, but they do not have to be viable to produce allergenic effects in sensitive people. Thus, although a bright sunny afternoon might produce enough ultraviolet light to substantially reduce the viability of fungal spores in the air, it will not bring relief to patients suffering from fungal allergy, while a rain storm will wash down fungal spores from the air, producing at least temporary relief for fungal allergy sufferers.

SURVEYS OF AIRBORNE FUNGI

Research on airborne fungi has been performed worldwide over the last five decades. Storm Van Leeuwen (82) was the first to suspect that airborne fungi played a role in patients' allergy attacks. He found fungi in the home and in the mattresses of a patient in Holland who suffered from asthma, and he suspected the fungi to be the causative agents. Others in Germany (31) and in the United States (10,34) studied airborne fungi as the causative agents of asthma and hay fever in many patients. Towey et al. (83) have reported an asthma incident in a patient who was exposed to fungal spores from the bark of maple logs, and Cobe (13) has described a similar case in which a greenhouse worker suffered an asthma attack after inhaling spores from *Cladosporium (Hormodendrum)* grown on a tomato plant.

Due to increasing awareness of the relationship of airborne fungi to allergy in patients suffering from asthma and hay fever, many scientists and allergists began to study the presence and type of fungal spores in both indoor and outdoor air, using various collection methods. The most common procedure involves the use of microscope slides coated with petroleum jelly or silicone that are exposed to the air for 24 hours. The fungal spores collected (as well as the pollens) are counted by the aid of a microscope. To aid in identification of fungi, the culture plate exposure method (gravity method) was adopted routinely to study the number and type of fungi in the air. These methods, as well as those that are more sophisticated (using specialized devices), are discussed in Chapter 4.

Reports and surveys of airborne fungi from all parts of the world now appear in the literature. These include reports from England (20,62,64), Ireland (52), Spain (12), Denmark (50), Finland (53), Greece (68), India (70,73), Iraq (6), Israel (7), Thailand (58), Nigeria (15), South Africa (60), Central and South America (57), Bermuda (14), Canada (8), Brazil (48), Australia (22), New Zealand (15), and various parts of the United States (1–5,18,19,25,35,56,71,72,74,75,77,78). There have also been surveys of greater specificity, such as those for air near the ground (26), at high altitudes (87), indoor-outdoor comparisons (37,38), inside certain quarters or buildings (21,49,59,81), or concerning certain groups of fungi, pathogenic, allergenic, or otherwise (23,51,84).

One of the most thorough studies of airborne fungi was performed in Kansas and detailed in a series of articles published under the title of "Kansas aeromycology" (39–47,63–66,71). The investigators in this series began by examining suitable media for cultivation and identification of isolated fungi (71). They tested six media under the same conditions and chose the rose bengal-streptomycin medium because of sufficient spore production and restricted hyphal growth. On this medium colonies remain discrete and do not overgrow each other, which aids in identification. The investigators continued to standardize materials and methods (39). This was followed by studying each genus or group of fungi in a separate investigation: *Cladosporium* (42), *Alternaria* (43), *Penicillium* and *Aspergillus* (44), hyphal fragments (63), smuts (64), Phycomycetes (45), Ascomycetes (40), Basidiomycetes (65), and imperfect fungi (41).

In my opinion, although these investigators' collection methods involved the old culture plate exposure method and surface-coated slides instead of the advanced sampler types presently available, the study still remains as the most organized, best documented, and well planned of airborne fungal studies.

Any survey that endeavors to obtain significant data should be repeated consecutively for at least 3 years to monitor the prevalence of the fungi found, and to determine both their quantitative and qualitative distribution. A one-season survey (whether for a few months or for one year) will only reveal that during that period of time certain fungi were found in a specific amount; it will not indicate whether the same data would be obtained again, within the same pattern, for the next year or the year following. Knowledge of the pattern of occurrence of the fungi helps the allergist in working with patients.

Certain essential factors should be established and evaluated prior to beginning any airborne fungal survey so as to obtain meaningful results. These basic factors remain the same throughout, even though investigators in different locations may use various approaches in their studies. These factors include the following:

1. What is the reason for the survey? Will it benefit allergists, general mycologists, or plant pathologists?
2. What fungi are we looking for? Are we looking for all types of fungi? for human pathogens? for plant pathogens? smuts and rusts? thermophilic Actinomycetes? the allergenic fungi?
3. The medium or media should be chosen according to the type of fungi being investigated. An appropriate medium will encourage fungus growth with enough sporulation for easy identification,

while at the same time it will discourage or eliminate bacterial growth or appearance of other undesirable fungi. A medium should be easy to maintain, and easy to handle. It is strongly recommended that more than one medium be used.

4. The collection procedure will depend on the type of fungi being sought. A thorough review of the techniques used in studying airborne fungi is recommended before a final decision is reached. In most instances more than one procedure is employed to ensure the complete recovery of all fungi needed.

5. The time of the day is crucial for collection of fungi. Early morning hours are unsatisfactory, because most fungi settle to the ground overnight. As daily activities increase, however, fungi are stirred up and the fungal population in the air increases, usually reaching its peak (under normal conditions) around 12 to 2 PM. It is important for the collection always to be done at the same time of day. Investigators should also take into consideration the diurnal periodicity of certain fungi and select collection times accordingly.

6. How often to collect depends on the preliminary planning of the study and the purpose of the data; for example, collection might be made daily, twice a day, twice a month, or twice a week. An alternative for collecting during bad weather should be included in the plan.

7. Location for collection should be decided. Choices range from ground level to above ground. How far should collection be above ground? top of the roof? near heavy traffic? done in a suburban area? in a downtown area? in front of the house or in the backyard? Whatever decision is made, the same site should be used throughout the study. Furthermore, open plates, when used, should be kept away from high walls or other objects that interrupt air flow.

8. The length of time the plate is to be left open or the air sampler is to be left in operation must be determined by several trials prior to beginning the survey. The goal is to have a sufficient number of fungi collected to provide significant data.

9. Surveys made inside the homes of allergenic patients should include the bedroom where the patient sleeps, because the patient spends most time here throughout a 24-hour period. The second most important place to sample is where the patient spends most waking hours, such as in an office in case of workers, in the basement if used by a child as a playroom, in the kitchen if the housewife spends most of the time cooking, and so forth. More than one location should be tested.

10. Note that certain fungi growing on agar plates form sister colonies, which may yield false counts. Any fungal colony that produces a large number of spores that disperse easily can create such a problem; the best example is *Penicillium*. One *Penicillium* colony growing

on a plate can produce spores that disperse to the agar surface and develop many new colonies. Observation of the size of the colonies of the same fungus may indicate formation of sister colonies.

11. A survey of only one season is insignificant, as mentioned previously, for atmospheric studies. Within a house or an office, surveys of from 1 to 3 days may yield significant data to help solve a problem for the allergist. For an atmospheric survey, however, 3 years of consecutive surveying is highly recommended to obtain significant data regarding types of fungi, increases and decreases in counts of fungi, peak periods, and correlation of such findings with climatic conditions to help define a pattern, so that allergists can know what to expect (to a certain degree). This type of study provides the most valuable information for allergists, and is the most helpful indirectly for the patient. In some cases patients may be strongly advised to stay indoors during certain times of the year or to avoid outdoor areas in which fungal spores are expected to be found abundantly. Three-year surveys should be repeated every 5 years to keep up with changes in climatic conditions and changes in surrounding vegetation that influence the pattern of airborne fungi. I would like to emphasize here once again, that only one survey a season or a one-year survey is almost useless for allergists in treating their patients.

Obviously, because the use of slides coated with petroleum jelly or silicone is of limited advantage for studying airborne fungi, they are unsuitable for qualitative studies of such fungi; identification is almost impossible in certain cases (Fig. 3–1).

The culture plate exposure method (gravity), which is the most common method used for surveying the types of airborne fungi in any vicinity, is by itself not entirely satisfactory for the study of such fungi because the procedure depends on the random deposit of the fungal spores from the air onto the surface of the agar in the plate. Therefore, the location of the plate, direction of the wind, and gravity of the various fungal spores are important factors that determine which fungi are captured. Furthermore, any artificial medium must of necessity be selective in that it permits the growth of some but not other varieties of fungi. Not every airborne fungus grows on artificial medium, and some will thus be lost to colony counts. Other hazards include overgrowth of young, not yet identified, colonies by rapidly growing and spreading fungi (e.g., *Rhizopus* or *Mucor*) and, finally, awareness of the fact that some fungi may elude identification because of failure to sporulate due to the medium chosen or because they belong to the mycelia sterilia group. For a real quantitative (as well as qualitative) survey of fungi, the Anderson sampler is strongly recommended (Fig. 3–2). (See Chapter 4 for a more detailed discussion.)

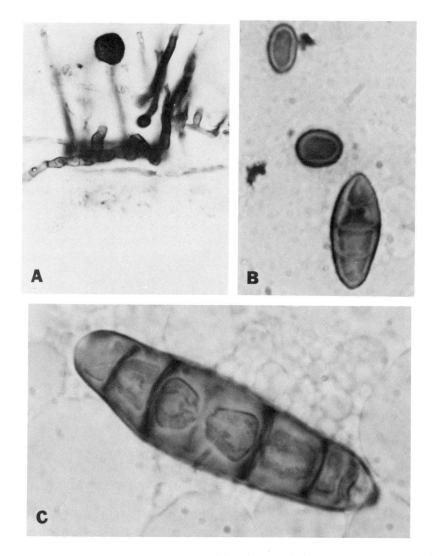

FUNGI PRESENT IN THE AIR

There are multi-thousands of recognized species of fungi. They are found in soil, in water, on animals, on vegetation, in humans, and in almost every part of the environment. Most airborne fungi are found as spores and a few as hyphal fragments. The most common airborne fungi belong to the genera *Cladosporium (Hormodendrum), Alternaria, Aspergillus, Penicillium, Helminthosporium, Aureobasidium, Phoma, Nigrospora, Rhizopus, Mucor, Epicoccum, Stemphylium, Curvularia, Fusarium, Scopulariop-*

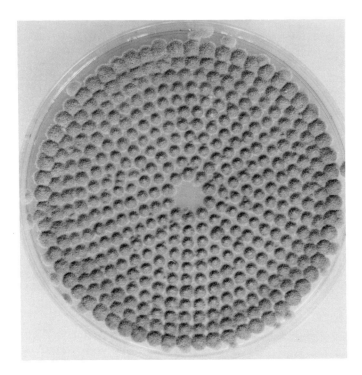

Fig. 3–2.   Plate from Anderson sampler (plate no. 6) using potato dextrose agar shows growth of colonies of *Aspergillus* sp. collected in 5 minutes of sampling (5 ft³ of air) at Washington, DC.

sis, Cephalasporium, Chaetomium, Trichoderma, Streptomyces, Candida, Cryptococcus, and *Rhodotorula,* as well as rusts, smuts, and hyphal fragments that could belong to mycelia sterilia, or other sporulating fungi. The spores of some of these fungi can be easily recognized on the exposed slides, but most cannot be identified except by their isolation on the exposed agar plates (Fig. 3–3). Occasionally, however, a fungus will be isolated that is not routinely obtained from the air (Fig. 3–4).

All fungi that cause allergies are airborne fungi. Most allergists believe that, in addition to the thermophilic Actinomycetes, there are about 40 to 45 fungi known to cause allergy in sensitive human patients. However, I have come to believe that an individual can develop an allergy to certain fungi for which he had not previously displayed an allergic reaction, even if some of these fungi may not have been known previously to cause allergy in humans. Within the field of medical mycology we believe that all fungi can be pathogenic to humans or animals if the fungi are capable of growing at 37° C and can find a suitable host. It is unfortunate that a large pool of suitable hosts for such opportunistic fungi is being provided due to a lack of control over most recent advances

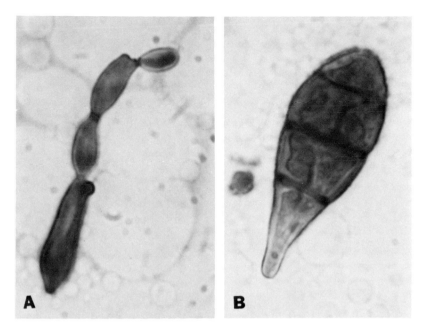

Fig. 3–3. Fungi collected on exposed slides show conidiophore and conidia of *Cladosporium* sp. *(A)* and a conidium of *Alternaria (B)*.

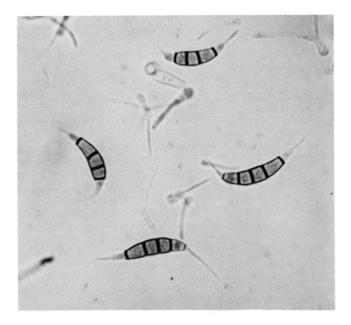

Fig. 3–4. Wet mount of spores of the fungus *Monochaeta* sp. isolated from exposed plate in Washington, DC.

in our medical technologies. This principle might apply in the field of allergy: all fungi can be allergenic if they find the "suitable sensitive" individual.

Cladosporium is the most common fungus found in the air, followed by *Alternaria, Penicillium, Aspergillus, Fusarium,* and *Aureobasidium pullulans* (Table 3–1). Unfortunately, *Cladosporium* and *Alternaria* are clinically the causative allergenic agents for most patients sensitive to fungi.

DIURNAL PERIODICITY

Fungal spores are always present in the air, with rain and snow washing down most if not all spores from the air, and sunshine and wind causing an increase in the atmospheric distribution of such spores. The number of airborne fungi is lowest during the winter months and highest during the spring and summer months. Certain fungi (spores), however, display a special increase or decrease at certain hours of the day or night, regardless of the usual climatic conditions. Such an increase (peak) or decrease of these fungi in the air from day to day or even from hour to hour during the day or night is known as a diurnal rhythm pattern. For

TABLE 3–1. Most Common Airborne Allergenic Fungi From 19 Random Surveys

		Percentages Within Total Counts				
Alternaria	Aspergillus	Aureobasidium pullulans	Fusarium	Cladosporium	Penicillium	Reference No.
30	6	0	13	27	4	73
28	2	0	2	39	3	2
17	1	7	2	19	6	22
35	10	15	2	14	7	25
2	12	0	14	27	6	58
34	9	15	2	14	7	19
20	1	0	0.1	40	3	75
22	6	0	4	32	5	3
12	7	4	10	30	19	5
16	2	19	1	21	6	4
2	6	1	0	51	9	74
0	15	4	24	0	21	46
13	5	1	1	45	6	39
0	0	21	28	0	0	45
12	13	0	2	35	17	71
0	2	0	8	37	2	16
24	2	4	0.2	28	11	35
9	3	0	1	23	11	60
9	3	0	0.1	69	6	50
14.0	6.1	4.7	5.6	29.2	8.8	Average

some reason, certain fungi show diurnal rhythms at all times, some show it at certain times only, and others do not seem to have this characteristic. Furthermore, some fungi display such peaks and valleys in distribution only in certain localities, and fail to do so elsewhere. This would seem to indicate that the diurnal periodicity of these fungi correlates with both climatic conditions in the area as well as with the appearance of certain vegetation or other environmental factors.

Studies of diurnal periodicity in airborne fungi require use of continuous-type samplers, which permit continuous sampling through a 24-hour period or longer. Many of these have been developed in recent years (32,61,67). Using these samplers, studies were undertaken by many workers where fungi were found to have peaks at specific times of the day or night (26–29,33,80,85). Some peaked in the early morning (85), some peaked just after midday (33), and a few peaked after rainfall (30). For example, Panzer et al. (67) found that *Cladosporium* and *Epicoccum* display a nighttime peak, while *Alternaria, Nigrospora, Penicillium*, and *Curvularia* have a daytime peak. Similarly, Pady (61) found that *Alternaria, Helminthosporium,* and hyphal fragments had daytime peaks between 3 and 5 PM. Barkai-Golan et al. (7) exposed nutrient plates for 48 hours and found that colonies of *Cladosporium* experienced a morning peak on one day and an afternoon peak the next, while *Alternaria* had only morning peaks. At the same time it was noted that both *Aspergillus* and *Penicillium* showed inconsistent patterns; *Alternaria* showed a 3 to 5 PM. peak as seen by Pady (Kansas) (61), but an early morning peak, as seen by Barkai-Golan et al. (Israel) (7). Such differences, as shown by the same fungus in different areas, are not unusual and may indicate that the environment rather than the fungal nature determines the diurnal periodicity.

In further studies by Pady et al. in Kansas (66) on diurnal periodicity in airborne fungi, it was found that most fungi did not show any such rhythm. However, *Fusarium*, yeasts, basidiospores, and hyphal fragments showed a nighttime peak while smut spores showed a daytime peak. In this study *Alternaria* showed no peaks while *Cladosporium* showed peaks sometimes during the day and at other times at night, with no consistency.

It would be greatly beneficial to both allergists and patients suffering from fungal allergy if all allergenic fungi showed definite diurnal periodicity patterns, and if such patterns could then be recorded for every location (such as a town or city). Patients would thus know in advance the fungal peaks and would be able to avoid overexposure to environments in which such fungi are expected to be found in large quantities . . . it is just a hopeful thought.

REFERENCES

1. Al-Doory, Y.: The fungal flora of the air near the ground in San Antonio, Texas. Mycopathol., **32**:313, 1967.

2. Al-Doory, Y.: Further studies of the fungal flora of the air in San Antonio, Texas. J. Allergy, **30**:145, 1967.
3. Al-Doory, Y.: Application of Anderson sampler in studying airborne fungi in San Antonio, Texas. Mycopathol., **42**:293, 1970.
4. Al-Doory, Y., et al.: Airborne fungi and pollens of the Washington, D.C., metropolitan area. Ann. Allergy, **45**:360, 1980.
5. Al-Doory, Y., Domson, J.F., and Best, J.: Further studies of the airborne fungi and pollens of the Washington, D.C., metropolitan area. Ann. Allergy, **49**:265, 1982.
6. Al-Tikriti, S.K., Al-Salihi, M., and Gailiard, G.E.: Pollen and mold survey of Baghdad, Iraq. Ann. Allergy, **45**:97, 1980.
7. Barkai-Golan, R., et al.: Atmospheric fungi in the desert town of Arad and in the coastal plain of Israel. Ann. Allergy, **38**:270, 1977.
8. Bassett, J.: An atlas of airborne pollen grains and common fungus spores of Canada. Research Branch of the Canada Department of Agriculture, Monograph No. 18, 1978.
9. Benjamin, R.K.: Merosporangiferous Mucorales. Aliso, **4**:321, 1959.
10. Benton, H.S.: Asthma due to a mold—*Aspergillus fumigatus.* JAMA, **95**:189, 1930.
11. Cain, R.F.: Studies of coprophilous Sphaeriales in Ontario. Univ. Toronto Studies Biol. Ser., **38**:1, 1934.
12. Calvo, A., et al.: Air-borne fungi in the air of Barcelona (Spain). III. The genus *Aspergillus* link. Mycopathol., **71**:41, 1980.
13. Cobe, H.M.: Asthma due to a mold; hypersensitivity due to *Cladosporium fulvum* Cooke. A case report. J. Allergy, **3**:389, 1932.
14. Dewdney, J.M., et al.: A clinical and environmental study of the aeroallergens of the islands of Bermuda. Clin. Allergy, **8**:445, 1978.
15. DiMenna, M.E.: A quantitative survey of aerial moulds and yeast in Dunedin, New Zealand. Trans. Br. Mycol. Soc., **38**:119, 1955.
16. Dransfield, M.: The fungal air-spora at Samaru, Northern Nigeria. Trans. Br. Mycol. Soc., **49**:121, 1966.
17. Duddington, C.L.: The Friendly Fungi. London, Faber & Faber, 1957.
18. Dupont, E.M., et al.: A survey of the airborne fungi in the Albuquerque, New Mexico, metropolitan area. J. Allergy, **39**:238, 1967.
19. Dworin, M.: A study of atmospheric mold spores in Tucson, Arizona. Ann. Allergy, **24**:31, 1966.
20. Evans, H.C.: Thermophilous fungi isolated from the air. Trans. Br. Mycol. Soc., **59**:516, 1972.
21. Ford, C.R., Peterson, D.E., and Mitchell, C.R.: Microbiological studies of air in the operating room. J. Surg. Res., **7**:376, 1967.
22. Frey, D., and Durie, E.B.: The incidence of air-borne fungi in Sydney. Mycopathol., **13**:93, 1960.
23. Friedman, L., et al.: The isolation of dermatophytes from the air. J. Invest. Dermatol., **35**:3, 1960.
24. Gilman, J.C.: A Manual of Soil Fungi. 2nd Ed. Ames, Iowa, Iowa State Univ. Press, 1957.
25. Goodman, D.H., et al.: A study of airborne fungi in the Phoenix, Arizona, metropolitan area. J. Allergy, **38**:56, 1966.
26. Gregory, P.H.: Spore content of the atmosphere near the ground. Nature, **170**:475, 1952.
27. Gregory, P.H., and Hirst, J.M.: Possible role of basidiospores as air-borne allergens. Nature, **170**:414, 1952.
28. Gregory, P.H., and Hirst, J.M.: The summer air-spora at Rothamsted in 1952. J. Gen. Microbiol., **17**:135, 1957.
29. Gregory, P.H., and Sreeramulu, T.: Air spora of an estuary. Trans. Br. Mycol. Soc., **41**:145, 1958.
30. Gregory, P.H., and Stedman, O.J.: Spore dispersal in *Ophiobolus graminis* and other fungi of cereal root rots. Trans. Br. Mycol. Soc., **41**:449, 1958.
31. Hansen, K.: Über Schimmelpilz: Asthma. Verh. Dtsch. Ges. Inn. Med., **40**:204, 1928.
32. Hirst, J.M.: An automatic volumetric spore trap. Ann. Appl. Biol., **30**:257, 1952.
33. Hirst, J.M.: Changes in atmospheric spore content: Diurnal periodicity and the effects of weather. Trans. Br. Mycol. Soc., **36**:375, 1953.

34. Hopkins, J.G., Benham, R., and Kesten, B.M.: Asthma due to a fungus *Alternaria*. JAMA, **94**:6, 1930.
35. Hotchkiss, M., Scherago, M., and Caplin, I.: A weekly mold survey of the air in Indianapolis, Indiana. Ann. Allergy, **21**:563, 1963.
36. Hudson, H.J.: Aspergilli in the air-spora at Cambridge. Trans. Br. Mycol. Soc., **52**:153, 1969.
37. Kozak, P.P., Jr., et al.: Currently available methods for home mold surveys. I. Description of techniques. Ann. Allergy, **45**:85, 1980.
38. Kozak, P.P., Jr., et al.: Currently available methods for home mold surveys. II. Examples of problem homes surveyed. Ann. Allergy, **45**:167, 1980.
39. Kramer, C.L., et al.: Kansas aeromycology. II. Materials, methods and general results. Trans. Kans. Acad. Sci., **62**:184, 1959.
40. Kramer, C.L., and Pady, S.M.: Kansas aeromycology. IX. Ascomycetes. Trans. Kans. Acad. Sci., **63**:53, 1960.
41. Kramer, C.L., and Pady, S.M.: Kansas aeromycology. XI. Fungi imperfecti. Trans. Kans. Acad. Sci., **63**:228, 1960.
42. Kramer, C.L., Pady, S.M., and Rogerson, C.T.: Kansas aeromycology. III. *Cladosporium*. Trans. Kans. Acad. Sci., **62**:200, 1959.
43. Kramer, C.L., Pady, S.M., and Rogerson, C.T.: Kansas aeromycology. IV. *Alternaria*. Trans. Kans. Acad. Sci., **62**:252, 1959.
44. Kramer, C.L., Pady, S.M., and Rogerson, C.T.: Kansas aeromycology. V. *Penicillium* and *Aspergillus*. Mycologia, **52**:545, 1960.
45. Kramer, C.L., Pady, S.M., and Rogerson, C.T.: Kansas aeromycology. VIII. Phycomycetes. Trans. Kans. Acad. Sci., **63**:19, 1960.
46. Kramer, C.L., Pady, S.M., and Wiley, B.J.: Kansas aeromycology. XIII. Diurnal studies 1959–60. Mycologia, **55**:380, 1963.
47. Kramer, C.L., Pady, S.M., and Wiley, B.J.: Kansas aeromycology. XIV: Diurnal studies 1961–1962. Trans. Kans. Acad. Sci., **67**:442, 1964.
48. Lacaz, C.S., et al.: Fungos anemofilos mas cidaded de Sao Paulo e Santos (Brasil). Rev. Hosp. Clin. Fac. Med. Sao Paulo, **13**:187, 1958.
49. Lacey, J., and Lacey, M.E.: Spore concentrations in the air of farm buildings. Trans. Br. Mycol. Soc., **47**:547, 1964.
50. Larsen, L.S.: A three-year-survey of microfungi in the air of Copenhagen 1977–1979. Allergy, **36**:15, 1981.
51. Lurie, H.I., and Way, M.: The isolation of dermatophytes from the atmosphere of caves. Mycologia, **49**:178, 1957.
52. McDonald, M.S., and O'Driscoll, B.J.: Aerobiological studies based in Galway. A comparison of pollen and spore counts over two seasons of widely differing weather conditions. Clin. Allergy, **10**:211, 1980.
53. Makinen, Y., and Ollikainen, P.: Diurnal and seasonal variations in the spore composition in Turku. S. Finland. Bull. Ecol. Res. Comm., **18**:143, 1973.
54. Malone, J.P., and Muskett, A.E.: Seed-borne fungi. Proc. Int. Seed Testing Assoc., **29**:177, 1964.
55. Middleton, J.T.: The taxanomy, host range and geographic distribution of the genus *Pythium*. Mem. Torr. Bot. Club, **20**:1, 1943.
56. Morrow, M.B., Meyer, G.H., and Prince, H.E.: A summary of airborne mold surveys. Ann. Allergy, **22**:575, 1964.
57. Naranjo, P.: Etiological agents of respiratory allergy in tropical countries of Central and South America. J. Allergy, **29**:362, 1958.
58. Nissaisorakarn, M.: Air-borne fungi in Bangkok and Thonburi. Proc. of the 18th Intern. Cong. Military Medicine, Sala Santitham, Thailand, 1965.
59. Noble, W.C., and Clayton, Y.M.: Fungi in the air of hospital wards. J. Gen. Microbiol., **32**:397, 1963.
60. Ordman, D.: The air-borne fungi in Johannesburg. A second five-year survey: 1955–1959. S. Afr. Med. J., **31**:325, 1963.
61. Pady, S.M.: A continuous spore sampler. Phytopathol., **49**:757, 1959.
62. Pady, S.M., and Gregory, P.H.: Numbers and viability of airborne hyphal fragments in England. Trans. Br. Mycol. Soc., **46**:609, 1963.
63. Pady, S.M., and Kramer, C.L.: Kansas aeromycology. VI. Hyphal fragments. Mycologia, **52**:681, 1960.

64. Pady, S.M.: and Kramer, C.L.: Kansas aeromycology. VII. Smuts. Phytopathol., **50**:332, 1960.
65. Pady, S.M., and Kramer, C.L.: Kansas aeromycology. X. Basidiomycetes. Trans. Kans. Acad. Sci., **63**:125, 1960.
66. Pady, S.M., Kramer, C.L., and Wiley, B.J.: Kansas aeromycology. XII. Materials, methods, and general results of diurnal studies 1959–1960. Mycologia, **54**:168, 1962.
67. Panzer, J.D., Tullis, E.C., and Van Arsdel, E.P.: A simple 24-hour slide spore collector. Phytopathol., **47**:512, 1957.
68. Papavassilious, J.T., and Bartzokas, C.A.: The atmospheric fungal flora of the Athens metropolitan area. Mycopathol., **57**:31, 1975.
69. Phaff, H.J., Miller, M.W., and Cooke, W.B.: The Life of Yeasts; Their Nature, Activity, Ecology, and Relation to Mankind. Cambridge, Harvard University Press, 1966.
70. Rajan, B.S.V., Nigam, S.S., and Shukla, R.K.: A study of the atmospheric fungal flora at Kanpur. Proc. Ind. Acad. Sci., **35**:33, 1952.
71. Rogerson, C.T.: Kansas aeromycology. I. Comparison of media. Trans. Kans. Acad. Sci., **61**:155, 1958.
72. Roth, A., and Durham, O.: Pollen and mold survey in Hawaii, July, 1963, to June, 1964. J. Allergy, **36**:186, 1965.
73. Sandhu, D.K., Shivpuri, D.N., and Sandhu, R.S.: Studies on the air-borne fungal spores in Delhi. Their role in respiratory allergy. Ann. Allergy, **22**:374, 1964.
74. Sayer, W.J., Shean, D.B., and Ghosseiri, J.: Estimation of airborne fungal flora by the Anderson sampler versus the gravity settling culture plate. I. Isolation frequency and numbers of colonies. J. Allergy, **44**:214, 1969.
75. Shapiro, R.S., Eisenberg, B.C., and Binder, W.: Airborne fungi in Los Angeles, California. J. Allergy, **36**:472, 1965.
76. Siu, R.G.H.: Microbial decomposition of cellulose. New York, Reinhold Publishing, 1951.
77. Sneller, M.R., Hayes, H.D., and Pinnas, J.L.: Frequency of airborne *Alternaria* spores in Tucson, Arizona over a 20-year period. Ann. Allergy, **46**:30, 1981.
78. Sorenson, W.G., Bulmer, G.S., and Criep, L.H.: Airborne fungi from five cities in the continental United States and Puerto Rico. Ann. Allergy, **33**:131, 1974.
79. Sparrow, R.K.: Aquatic Phycomycetes. 2nd Ed. Ann Arbor, University of Michigan Press, 1960.
80. Sreeramulu, T.: The diurnal and seasonal periodicity of spores of certain plant pathogens in the air. Trans. Br. Mycol. Soc., **42**:177, 1959.
81. Stallybrass, F.C.: A study of *Aspergillus* spores in the atmosphere of a modern mill. Br. J. Ind. Med., **18**:41, 1961.
82. Storm Van Leeuwen, W.: Weitere untersuchungen über asthma und klima. Klin. Wochenschr., **4**:1294, 1925.
83. Towey, J.W., Sweany, H.C., and Huron, W.H.: Severe bronchial asthma apparently due to fungus spores found in maple bark; with discussion. JAMA, **99**:453, 1932.
84. Treuhaft, M.W., and Arden Jones, M.P.: Comparison of methods for isolation and enumeration of thermophilic Actinomycetes from dust. J. Clin. Microbiol., **16**:995, 1982.
85. Waggoner, P.E., and Taylor, G.S.: Dissemination by atmospheric turbulence of spores of *Peronospora tabacina*. Phytopathol., **48**:46, 1958.
86. Wogan, G.N. (ed.): Mycotoxins in Foodstuffs. Cambridge, MIT Press, 1965.
87. Wolf, F.T.: The microbiology of the upper air. Bull. Torr. Bot. Club, **70**:1, 1943.

Chapter 4

SAMPLING TECHNIQUES FOR AIRBORNE FUNGI

William R. Solomon

Recognition of the relation of inhalant allergy to fungi was facilitated initially when symptoms were discovered among workers exposed to dense spore clouds while working with cereal crops (7). In most cases, however, allergic symptoms to fungi follow far more subtle and inapparent challenges from natural aerosols; often such exposure originates at substantial distances from the affected subjects.

Substrate surveys offer limited insight into the prevailing types and levels of airborne fungi. Unlike pollens, animal danders, or even house and commercial dusts, major fungus emanations tend to have sources that are microscopic; however, these are abundantly distributed in nature. In practice, culture-based field analyses regularly yield bewildering arrays of species, most of which contribute little to atmospheric loads. With no other reliable indicators of fungus prevalence, direct sampling of air continues to provide a uniquely valuable index of clinical exposure. Advances in our knowledge of fungi in air have quickly followed refinements in sampling methodology (16). Not surprisingly, many enduring misconceptions in this field may also be traced to deficient or misapplied collection techniques.

With few exceptions, solid particles have been the recognized carriers of allergenic activity, confining effort and interest to this aerosol class; even so, various recovery techniques continue to be applied to airborne fungi. In addition to reflecting the often traditional preferences of individuals and specific disciplines, this technical diversity underscores the lack of any universally "ideal" strategy of collection. Indeed, it is clear that each approach "selects" a characteristic spectrum of biogenic agents for investigative attention.

OBJECTIVES AND LIMITATIONS OF ATMOSPHERIC SAMPLING

Data reflecting the occurrence of fungus materials in air are sought by several disciplines, each with different informational needs. Plant pathologists frequently focus solely on viable particles and may be more concerned with deposition onto surfaces than with levels of suspended units. Because the advent of a transported pathogen, even in trace amounts, must often be sensed, relatively high sampling volumes and well-mixed outdoor air are sampling prerequisites. By contrast, fungal physiologists may require tiny sequential samples obtained within millimeters of infected leaf surfaces. With intramural outbreaks of human respiratory disease, "grids" of spot samples may help to map indoor sources of microbial growth and routes of transport. These differing needs demand goal-directed choices among sampling techniques as well as in the location and scheduling of collections. As a result, biogenic aerosol ("air spora") data, rigorously gathered for one purpose, often have limited adaptability to others.

The needs of clinical allergy are varied but generally emphasize indices of fungus prevalence that may be compared spatially and temporally as estimates of exposure risk. This application is fostered if data are "volumetric" (that is, validly related to unit volumes of air). A related requirement, of course, is for sampling devices that will harvest particles from defined volumes of air without bias. Even with ideal methods of collection, however, concepts of *individual* exposure must be developed with caution. Most prevalence data are traditionally derived from urban rooftop samples; these describe a regionally variable "background" of fungal challenge. Levels of fungus particles nearer the earth's surface are often substantially higher than at the rooftop and may be directly increased by human activity (34). Furthermore, local environmental sources determine striking qualitative differences among the air sporas of natural sites, and create indoor-outdoor disparities in exposure potential as well (39).

Because automated analytic procedures have not been adapted to mixed natural aerosols, particles must be deposited onto surfaces before their enumeration, microscopically or in culture. This initial processing step poses problems that reflect intrinsic characteristics of airborne particles as well as limitations imposed by atmospheric motion. For any particle type, the presence or lack of a morphologically distinctive appearance probably most influences the choice of collection strategy. Agents unrecognizable by form alone must be either cultured or monitored by other methods (e.g., immunochemical analysis) that are under investigation. Clearly, many types (especially diverse ascospores and basidiospores) are not identifiable as specific particles *or* as sources of distinctive colonies on available media; they currently defy all enumerative efforts. For the small spherical spores typical of many genera of Deuteromycetes and Zygomycetes, assessing prevalence depends on the feasibility of culture. Particles "alive" by other criteria may fail to form colonies because of deficient substrates, unfavorable incubation conditions, or the presence of biogenic inhibitors of germination and hyphal growth. Furthermore, because substantial fractions of air spora appear to be "nonviable" by all criteria (25), most culture-based data yield far lower estimates of prevalence than would direct particle counts (5). For such common types as *Cladosporium* species the resulting differences in calculated numbers of particles may approach two logs; this has prompted increasing reliance on microscopic enumeration of aerosol units (5).

In the course of active discharge (as "ballistospores"), dry wind scouring, or splash dispersal, fungus particles enter a generally drier medium in which they are subject to sustained accelerations. Dispersed units appear to carry static charges that may affect their deposition onto natural and sampling surfaces; the extent of these effects is unknown (16). In addition, variations in moisture content may produce differences in the shape and size of dispersed particles from values typical of the same materials when observed in aqueous mounting media. Athough most determined values have been close to 1.0, relatively few estimates of density have been made for harvested fungal spores (16).

These uncertainties and the markedly nonspherical form of many fungus particles limit predictions of their behavior based on presumed physical properties. A coherent approach is made possible, however, through consideration of the "equivalent aerodynamic radius" of a specific aerosol type (i.e., the radius of the appropriately dimensioned, unit-density sphere that behaves in the same way as the agent being investigated). This attribute has been defined for relatively few particle types but may be inferred from experimental deposition patterns in samples of known size selectivity, such as cascade impactor recoveries. Subsequent references to "large" and "small" particles carry implications of equivalent aerodynamic radius for all categories (including spherical ones).

Probably the most persistent problems in fungal sampling may be traced to the remarkable diversity of airborne types, especially in regard to their broad range of size and shape. Because emanations of the various fungi differ widely in prevalence levels, errors of estimate, with any sample size, will also differ. Moreover, aerometric data often are found not to be distributed normally, making the use of nonparametric statistical methods essential (5). However, the need to collect fungus particles (often sized from below 2 to over 60 μm) with quantitative validity presents an overriding challenge. That challenge confronts the use of each of the various collection devices.

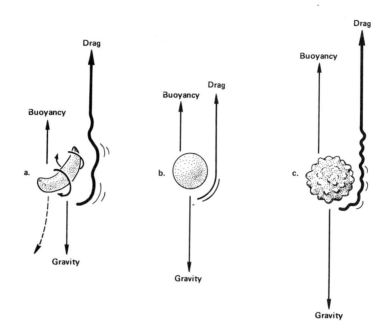

Fig. 4–1. Forces that act on particles during passive fallout in still air: gravitational acceleration per unit mass is constant, and buoyancy varies directly with volume. Drag is
minimal for an idealized smooth spherical unit (b) but increases with greater surface
roughness (c). The tumbling motion of an asymmetric particle (a) raises drag; in addition,
such aerosol species tend to "yaw" (deviate from strict verticality) during free fall, as the
dashed line indicates.

PARTICLE BEHAVIOR IN STILL AND MOVING AIR

Airborne particles follow complex paths that reflect "drag" (i.e., their
frictional interaction with adjacent air molecules) and buoyancy as well
as gravitational pull and other accelerations (Fig. 4–1). Comparatively
small units with low ratios of mass to surface area are especially influenced by drag forces, and usually follow the flow of entraining air closely.
The paths of larger aerosols are less affected by adjacent gas molecules
and tend to show clearcut *linear* deflections after accelerations are applied. Consequently, as particle size increases, trajectories more readily
cross the lines of flow of surrounding air. These "inertial" effects are
exploited directly to promote deposition of suspended materials onto
collection surfaces.

Still air situations occur seldom in nature but offer insight into the
forces affecting aerosol behavior. If, for example, a particle is introduced
into still air, it will undergo gravitational acceleration against the op-
posing drag of air molecules and the particle's innate buoyancy (Fig.
4–1). As fall speed increases these vector forces will reach an equilibrium,
and the particle will continue toward the surface at a fixed rate. This
terminal velocity, or Stokes settling velocity (V_s), is an important deter-

minant of fallout potential; it can also predict behavior at vertically placed obstructions. For most microscopic fungal units, the settling velocity may be estimated with the following equation. Correction factors are necessary for particles larger than 50 μm. Even for 100 μm units, however, V_s is only *ca.* 15% above the calculated value.

$$V_S = \frac{D^2 g}{18 \mu} (\pi_{part} - \pi_{air})$$

where D is particle diameter; g is 980 cm/sec^2, π_{part} is particle density, and μ and π_{air}, respectively, are the viscosity and density of the still air involved. From this relationship (Stokes' law) the predominant influence of particle size on fallout potential is evident. In theory, a spherical agent of 20 μm diameter will fall out 100 times more readily than an agent of 2 μm, assuming equivalent densities and surface characteristics (Fig. 4–1). Because relatively few estimates of terminal velocity have been made for specific spore types, this formulation often must remain untested. However, available data show a range of values exceeding two orders of magnitude that accords generally with overall particle size (16).

If an instantaneous horizontal acceleration is applied to a particle, it will travel a path of characteristic length before being brought to rest by air resistance. This "stop distance" is proportional to the unit's initial velocity and to its V_s, being the product of V_s/g and velocity for very small and/or slow-moving aerosols. For larger and/or more rapidly moving particles, this relationship must be adjusted to reflect appropriate drag forces, and values may be derived from suitable nomograms. (15,31). In general, however, with increasing particle size, stop distance increases exponentially, although the rate of increment is greater at lower than at higher projection velocities. Aerosols with relatively large stop distances do not change direction readily as an obstacle is approached and tend to impact on suitably positioned collection surfaces. By contrast, smaller units with very much shorter stop distances are more readily redirected by air flow around obstructions, thereby tending to avoid interception. Smaller particles will also change direction to enter a suction device far more easily than larger (or more rapidly moving) aerosol types. The concept and influence of stop distance are best visualized in still air models. Yet, effects of aerosol size and speed are equally significant when moving air streams supply continual accelerations to entrained particles.

STRATEGIES OF COLLECTION

Efforts to collect fungus particles for enumeration microscopically or in culture have developed over more than 150 years; techniques have become increasingly innovative recently. Essentially all contemporary methods use one of four major sampling approaches: fallout, impaction on narrow surfaces, filtration, and acceleration with impingement of particles at flow channel bends.

Gravitational Fallout

Sampling by passive fallout on slides or culture media was explored by workers as early as Pasteur and Miquel who were, doubtless, intrigued by its mechanical simplicity; however, both judged the method inadequate for quantitative recovery of bacteria (16). Decades later, Blackley exposed horizontal slides and observed collected fungus spores and pollen. Use of fallout for sampling airborne allergens was greatly stimulated by Durham's proposal in 1946 of a standard "gravity slide" method (13). This approach uses two horizontal polished Monel metal discs, of 9-inch diameter, separated by three 4-inch-long struts (Wilkins-Anderson Company, Chicago). A support for a 1 × 3 inch microslide is mounted 1 inch above the center of the lower (support) disc; the upper disc serves as a rain shield. Mounting of the device on a pipe or ring stand is facilitated by a threaded flange or by a collar and set screw on its underside. In practice, slides have been coated with a hydrophilic adhesive such as soft glycerin jelly (prepared by mixing 5 g of gelatin, 4 g of phenol, 40 ml of glycerin and 155 ml of water, and warming gently) or a nonwettable grease, and exposed for 24-hour periods. Usually, only a central square with 1.0 or 2.2 cm sides is scanned (and particles per centimeter square derived arithmetically), despite the relative sparsity of deposition. By removing the slide support, exposure of open plates of culture medium in the "Durham shelter" has been attempted; with the upper plate serving as a baffle, however, the device is not well suited to this application.

"Gravity" collectors are simple, independent of power sources, and largely maintenance-free. Furthermore, the recovery of airborne agents, often in patterns that roughly parallel clinical events, has fostered widespread confidence in the validity of fallout techniques. This intuitive endorsement has been challenged increasingly, however, in recent years. The inability to determine the volume of air that contributes particles for any sample is a fundamental limitation of passive fallout. Results of studies with particulate markers have confirmed that deposition onto horizontal slides varies with the speed, direction, and turbulence structure of the entraining air (35). Without a volumetric common denominator, therefore, it is impossible to assume differences in prevalence from differences in recovery as long as atmospheric conditions vary among collection periods. In a normally moving atmosphere much deposition is supplied by eddies inherent in turbulent flow or created as the slide margin perturbs oncoming flow lines. Similar factors affect collection by open culture dishes with even greater eddy information anticipated at the 10 to 15 cm projecting walls (Fig 4–2). As shown, these obstructions tend to create deposition "shadows" that vary with air speed and turbulence levels and further reduce particle recoveries.

Figure 4–2 also depicts the principal drawback of gravitational collection: that is, the tendency for recoveries to be strongly biased in favor of relatively large aerosol particles. This pervasive trend reflects settling velocity as an exponential function of size as well as the similar relationship that determines stop distance. Numerous comparative studies have confirmed the size-dependent bias inherent in collections by gravity slides and open culture plates (41, 44), leading to the virtual exclusion of small micronic types. In practice, the appearance of at least some aspergilli and other small-spored types on open culture plates has made it difficult, especially for physicians, to acknowledge the quantitative deficiencies of this approach. When examined carefully, however, recoveries have often been found to be unrelated numerically to actual prevailing levels of these colony-forming units (44). Deposition, especially among Deuteromycetes, may be fostered by the dispersion of tiny spores in clumps or chains (creating an aerodynamically larger unit) or by the "rafting" of hyphae or spores on other particles. Because fungus particles encompass such a broad size range, with many relatively minute forms, present fallout data provide poor estimates of absolute *and* relative prevalence for these diverse forms. In addition, because gravi-

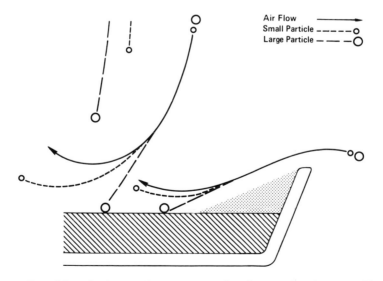

Air Flow ————⟶
Small Particle – – – – –o
Large Particle — — —O

Fig. 4–2. Deposition of "air spora" on an agar surface by a moving air stream. The relationship of increasing particle size to progressively greater fallout potential is suggested at the left. Toward the right a moving air stream deposits particles, from eddies, in accord with their stop distances so that, again, mainly larger units are recovered. With rapid air motion the lip of the culture plate creates a particle-free "shadow" *(stippled area)* over part of the collection surface.

tational methods typically provide rather small collections, sampling error is usually substantial, creating a situation in which less abundant types of spores may be entirely absent from the sample.

The use of a square gravity slide or wind orientation of the Durham shelter has been proposed to standardize the "effective deposition area" by continuously aligning air flow and slide axis. Higher recoveries have been reported by exposing a greased slide inclined at 30° and mounted on a wind vane (50); for each variation, however, the basic uncertainties of fallout collection remain. Where air flow is brisk, open culture plates poured to the height of the lip might offer advantages by mitigating "deposition shadow" effects; few trials of this modification have been evaluated, but its overall advantages presently appear to be limited.

Several additional attempts to promote passive deposition by employing aerodynamically formed adaptors deserve attention. Leuschner and Boehm have evaluated a "personal" slide sampler and noted distinctive recovery spectra apparently reflecting individual activity patterns (26); as with previous approaches (20), quantitative inferences were not possible. The Aerotest sampler (Olympic Medical, Seattle) is designed to detect contamination of respirator effluents and, in forced air streams, can distinguish high- from low-risk aerosols (40). Despite this useful application, there is no reason to expect comparable performance in still air in which simple open plates appear to offer higher recovery potential.

Impaction on Narrow Surfaces

The rapid motion of air in relation to narrow obstructions can provide a highly efficient mechanism for particle collection. This approach has been exploited using various slender manmade discs, ribbons and rods as well as the faces and edges of microslides (31, 36, 38). Analyses by several workers have indicated that the probability of impaction varies with the terminal velocity (V_s) and relative air speed of candidate particles (i.e., with their stop distances), and inversely with the width of the target (16,18,31). Actual recovery of aerosol units requires, of course, that the impaction event be followed by sustained adherence to an intercepting surface.

A stationary obstacle will capture specific moving particles with calculable efficiency if the speed of the entraining air is measured. This principle has been employed in the "flag" sampler, a 5-cm-long pin bearing a transversely applied segment of transparent tape (18). The trailing tape ends adhere behind the pin, forming a small wind vane, and the pin is free to rotate in a fused glass bearing. The leading tape edge is greased and forms a collection surface that may be removed and mounted for microscopic examination. Volumetric potential is realized by operating a totalizing anemometer (wind speed gauge) during ex-

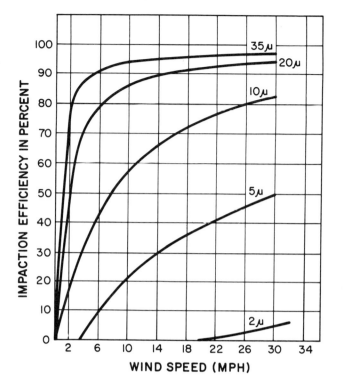

Fig. 4–3. Theoretic impaction efficiencies for unit density particles encountering a 1-mm wide surface in air moving at different speeds. The probability of impaction is quite low for the smallest unit (2 μm) and increases rapidly with particle diameter (and stop distance). (From Harrington, J.B., Gill, G.C., and Warr, B.R. [18])

posure periods. Using this readout the volume of air processed may be calculated as follows:

$$\text{Volume} = \text{average wind speed (cm/sec)} \times \text{time (sec)}$$
$$\times \text{sampling area (cm}^2\text{)}$$

where

Area = tape width (cm) × diameter of the tape-wound pin (cm)

Although it is most appropriate simply to relate particle counts to volume processed, relative estimates of impaction efficiency are possible, because this rises systematically with particle size and wind speed (18). Figure 4–3 displays theoretic values of impaction efficiency for several particle sizes over a range of wind speeds.

The flag sampler merits attention as a prototypic impaction collector but, at most, provides a means of plotting limited spatial concentration gradients. More widely useful impactors employ a sampling surface that

Fig. 4–4. AC-powered intermittently operating rotorod sampler. The small rain shield has been removed. One rod, in its support assembly, is shown extended at the right as it might appear during sampling intervals; its mate (not shown) is retracted into the crosspiece, where both remain between operating periods.

is whirled rapidly in a circular path to standardize air speed and to eliminate wind direction as a variable. The rotorod, developed by Perkins (Ted Brown Associates, Los Altos Hills, California), was the first such device to receive wide attention and, in several models, remains the only impactor now available commercially. Additional samplers, using microslide edges or narrow transparent tape-wound metal bars, have been more or less widely used, but lack current supply sources (18,36,38).

Rotorod devices use adhesive-coated clear lucite rods, 1.3 mm in width, as collection surfaces; the rods are whirled at 2400 rpm on suitable support assemblies. DC-powered models, designed for short periods of continuous sampling, are widely used; in addition, an "aeroallergen model" uses line voltage (Fig. 4–4). The latter unit incorporates a recycling timer that facilitates intermittent operation for 1 minute of every 10. (Although these samplers are supplied for 1-minute operating periods, this program is easily changed to allow intervals of from a few seconds to 5 minutes in each 10-minute cycle.) Silicone grease, as supplied by the manufacturer, is a preferred adhesive, performing at least as well as other lubricants in limited comparisons (46). By contrast, greaseless coatings, such as sticky tape or thinned rubber cement, rapidly lose their particle-retaining properties (18). Even excellent adhesives, however, show progressively decreasing efficiency during operation due to loading of the surface with deposited materials. The intensity of this effect does not correlate predictably with prevailing levels of biogenic particles but can be minimized by curtailing sampling time (43).

In the lower Great Lakes during summer, for example, an operating program of 15 to 20 seconds per 10-minute cycle (with a total of 36–48 minutes per 24 hours) has allowed the highest relative recoveries (43).

Following exposure, rods may be positioned in a Lucite stage support (available from the manufacturer) for microscopic examination. A cover slip of appropriate width is applied over an aqueous mounting medium, such as Calberla's solution (prepared by mixing 5 ml of glycerin, 10 ml of 95% ethanol, and 15 ml of distilled water; water-soluble dyes may be added, if desired). This arrangement provides favorable optical conditions, but semisolid preparations may be needed in rare instances in which oil immersion is required. Exposed portions of rods are examined (i.e., the distal 20 mm for I-type rods) and, in most cases, these areas are counted in full. Sampling volumes may be calculated for each rod as follows:

Volume sampled (cm³) = rod width (cm) × length of collection segment (cm) × diameter of rotation (cm) × π × rpm × minutes of sampling

Division of the product by 10^6 gives the sampling volume in cubic meters; recoveries are best expressed as "particles retained per cubic meter sampled" for comparative purposes. Collection with this method appears to vary little with ambient wind speed below a level of 15 mph, but may become less efficient at higher wind speeds. For any wind condition impaction efficiency correlates strongly with particle diameter, and for impactors, unlike gravitational devices, this capability may be estimated for any aerosol size fraction. Still, the practical value of impactor sampling for agents smaller than 5 to 8 μm in size is limited. Studies that focus on more minute spores have relied on the superior collection characteristics of suction devices for smaller particle fractions.

Filtration

The collection of particles by passing air through an extensively supported porous matrix is a relatively simple approach, proven valuable over a century ago. Generally, air is drawn through the filter by a pump or by some other negative pressure source for measured periods. Intercepted particles may be viewed microscopically by clearing the filters optically or, in some cases, by dissolving the filters and processing the resuspended materials. Alternatively, exposed filters may be floated on the surface of liquid culture media, which percolate upward through the minute channels and promote colonial development *in situ*.

Because filters and their supporting grids often impose significant resistance to flow, direct measurements of throughput, with all components connected in line, are essential. Flow readings are best made with a rotameter, of suitable range, connected by an airtight, low-resistance coupling immediately upstream of the filter assembly. When an in-line flow meter is to be used between sampler and suction source it should be calibrated, similarly, with a primary standard on the inflow side of the collection device.

A passive filtration unit, combining elements of impaction and filtration, has been described by French researchers for extended collections at remote sites (9). The device consists of two 20 × 20 cm matrices, each comprised of five thicknesses of silicone oil-soaked surgical gauze; these are wind-oriented on a large vane carried by a low-friction bearing. Particles are deposited by air passing through the overlapping layers at rates roughly dependent on wind speed. Following exposure the filters are digested in hydrofluoric acid, leaving deposited particles in suspension. At least some fungus spores survive this process morphologically intact but after-effects, especially on small ballistospores, are uncertain. The feasibility of culturing intact sterilized filters after exposure has been suggested but has not been tested. Experience with pollens indicates that very large collections may be obtained during week-long sampling periods, with extensive arrays of particle types recorded. The relative capture efficiencies of this device for different aerosol fractions, however, has not been defined under specified wind conditions. Furthermore, if data are to have any volumetric implications, a totalizing anemometer (and a power source to operate it) are essential. This requirement and the rather formidable processing steps required suggest that these filter units have marginal potential as samplers for fungi.

More conventional filter units, operating with defined throughputs, are widely useful in conditions in which particle loading is limited. Like all other suction devices, filter surfaces must be faithfully wind-oriented for all but extremely low-flow applications (18). Failure to maintain alignment of the intake with air flow results predictably in significant losses of efficiency, which worsen with increasing particle size and ambient air speed (Fig. 4–5). Under commonly encountered conditions, recoveries of spores (e.g., of *Alternaria* species) by samplers directed at right angles to air flow may show no more than a small percentage of the spores collected by a wind-oriented unit (47).

As shown also in Figure 4–5, inequalities in air speed between the filter inlet and the surrounding free stream may affect collection efficiency. In nature such differences normally wax and wane continuously and are extreme under gusty conditions; the resulting adverse effects on particle collection, however, generally do not approach those expected from directional misalignment. Unbiased sampling can occur when inlet and ambient air speeds are equal, so that flow lines enter the unit undeviated. This is a difficult condition to maintain but it is ideal, and is often termed "isokinetic." Deviations, or "anisokinetic" regimes, tend to distort recoveries in patterns that are dependent on particle size and air speed. When inlet velocity dips below ambient

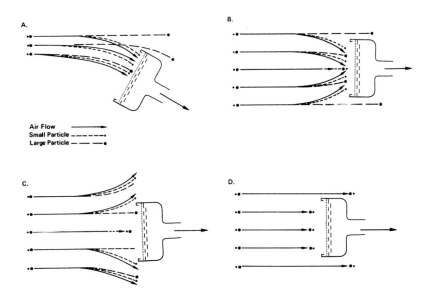

Fig. 4–5. Behavior of small and larger particles at the intake of a suction device. *A.* Orifice is misaligned with respect to wind direction. *B.* Intake is wind-oriented, but intake speed is well above that of the free stream. *C.* Intake is wind-oriented, but intake speed is well below ambient rate. *D.* Intake is wind-oriented, and speeds at intake and in free stream are equivalent (isokinetic). See text for additonal details.

velocity (Fig. 4–5) the entrance region will act as a relative stagnation point from which flow lines deviate outward. Quite small particles are apt to follow the entraining air; units with longer stop distances, though, tend, to proceed (in straight lines) into the orifice, causing factitiously high recoveries. Antithetical effects are expected when inlet air speed exceeds ambient air speed, so that flow lines converge to enter the orifice. Here again small particle paths comply closely with air flow and show relatively small losses. By contrast, large particles will tend to maintain "precommitted" directions, crossing streamlines at the intake and often escaping collection.

Limited studies have suggested that sampling errors resulting from these circumstances are lowest for aerosol units of roughly 20 μm and increase progressively as particle size extremes are approached (51). Isokinetic samplers have been designed for special applications (37); most suction collectors, however, including filter units, lack this capability. The resulting errors are minimized by maintaining wind orientation, by approximating inlet speeds to prevailing wind values, and by choosing orifices with relatively large areas. Suction collectors lacking wind orientation are best restricted to sites with essentially calm conditions. Fixed filter units, especially in undisturbed indoor air, may provide an economical volumetric option that has been largely overlooked

(10). Among the many filters and supports suitable for air sampling, those made by Gelman and Millipore have proven especially popular.

Suction Traps and Other Impingers

A wide variety of samplers use impingement as a collection strategy. This two-stage process involves aspiration of measured volumes of air into the device and raising this input and its entrained particles to critical velocities just before sharp bends in the flow channel are reached. Particles with stop distances exceeding available clearances will strike (i.e., "impinge" on) the walls at points suitable for collection surfaces. The probability of capture varies with particle size, acquired speed, and acuteness of the turn imposed, as well as with adhesive properties of the impingement point. Factors affecting the initial stage of this process are identical to those previously described for open filters. Air speeds at the entrance orifice and in the free stream should coincide as closely as possible and, except when conditions approach "calm," wind orientation is indicated. Intake characteristics of these collectors also determine that small particles will be captured with especially high efficiency. Furthermore, most suction devices produce accelerations sufficient to ensure the impingement of all or most aspirated micronic aerosols. Losses due to "slippage" (i.e., passage of particles entirely through the sampler) may be controlled with in-line back-up filters.

Liquid-Containing Impingers

These devices, which are among the simplest and least costly of volumetric collectors, provide particle suspension in small well-mixed fluid volumes. After filtration of the catch, particles may be viewed microscopically or may be cultured by flotation on semiliquid growth media. During operation with a suction source, aspirated air is accelerated by passage through a submerged constriction. This "nozzle" is positioned several millimeters from the flat bottom of the fluid reservoir or immediately above a disc suspended by struts from the narrowed inlet tube (Fig. 4–6). Aspirated air bubbles striking these surfaces are fragmented; entrained aerosol units will enter the fluid phase as air-fluid mixing occurs. Shear stresses incident to this process can be pronounced, causing separation of spore clusters and potentially damaging susceptible particles; fluid impingers, however, have been employed successfully for relatively delicate air spora. Because their inlet orifices lie in a horizontal plane, typical all-glass impingers are best adapted to calm air applications. Wind-responsive intakes for these devices have not appeared, although preimpingers, of defined efficiency, may be used to retain larger micronic fractions selectively (32). Liquid impingers are widely used for short-period, relatively low-volume sampling in enclosed spaces. Modifications in which impingement occurs on (rather

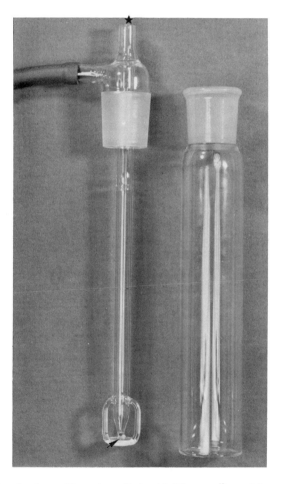

Fig. 4–6. All-glass impinger. The outer cylinder *(right)* normally contains a defined volume of fluid in which the inner tube *(left)* is partially submerged. The jet orifice *(arrow)* and impinger plate are closely opposed. A star marks the sampler intake.

than below) the fluid surface have been described, and may minimize trauma to delicate microorganisms (33).

Solid-Surface Collectors

Impingement on semisolid or on adhesive-coated solid surfaces has been exploited widely for sampling viable and nonviable particles, respectively. Furthermore, by choosing nozzles of diminishing size or by positioning collection surfaces increasingly closer to these "jet" orifices, the impingement of progressively smaller particle fractions is ensured. Cascade impactors employ a graded series, usually four, of such impingement stages to accelerate aerosols progressively (28). The result is a series of deposits with decreasing mean particle diameters, the largest size fraction appearing on the first collection surface encountered. In the most widely used version, collection is on greased glass discs and a back-up filter can be used to correct for slippage. Cascade impactors may be wind-oriented on suitable supports and, when isokinetic conditions are approximated, have provided a standard for other volumetric collectors. Because the 10 to 20 (often 17.5) liters/minute throughput impinges in narrow bands, however, collection surfaces readily become overloaded. Although the development of a model with moving collection surfaces largely overcame this problem (29), cascade impactors are only rarely used today for direct sampling of fungus aerosols. These devices are noteworthy as functional antecedents of the widely popular Hirst trap and of related suction samplers.

The automatic volumetric (Hirst) spore trap is an impinger with inlet dimensions of the second stage of the May cascade impactor (19). Air is drawn at a rate of 10 liters/minute through a 14 × 2 mm slit situated immediately proximal to the adhesive-coated collection surface. In its original form a 25 × 75 mm greased glass microslide provided the sampling area; a transparent tape-coated metal drum has been substituted in the contemporary Burkard (Hirst) trap (Fig. 4–7). Individual lubricants, including petrolatum, 90% petrolatum, and 10% mineral oil, silicone grease, and Lubriseal (A.H. Thomas, Philadelphia) have been applied as adhesives to the Melenex (transparent) tape supplied by the manufacturer. Of these coatings, Lubriseal has proven especially versatile, being compatible with mounting media as diverse as Permount (Fisher Scientific), hard glycerin jelly (made by adding 40 g of gelatin to batches of soft glycerin jelly, as previously described), and polyvinyl alcohol-based agents. A key-wound clock mechanism advances the sampling area past the intake slit at a rate of 2 mm/hour. Because the Burkard drum's circumference exceeds 340 mm, it easily accommodates 7 days of deposition between changes. Inflow is measured by a small rotameter adapted by a molded gasket to the aerodynamically flared entrance orifice. The intake rate may be easily adjusted by interposing a tube with a variable resistance side arm in the attachment to suction.

Fig. 4–7. Burkard trap operating at a rooftop site. *A.* The ample wind vane is evident, and the intake is marked *(arrow).* *B.* An exposed drum. Variations in particle recovery with time produced the banded deposit shown.

Following exposure, Burkard tapes are removed, laid flat and, usually, cut into strips of 48 mm length which represent 24 elapsed hours. These segments are affixed to glass slides using 10% Gelvatol (polyvinyl alcohol) aqueous solution or some other optically clear, semipermanent adhesive. Mounting of deposits, under 20 × 60 mm coverslips, has been accomplished with various media in strict accord with the adhesive employed. Especially satisfactory results have followed use of hard glycerin jelly or a Gelvatol mountant and sealing of the preparation with colorless nail lacquer.

Microscopic examination of spore trap deposits is best undertaken with "high-dry" optics or with an oil immersion lens having a "long" working distance. Longitudinal traverses provide "average" prevalence estimates for the collection period considered (i.e., overall particles counted per volume of air contributing particles to the strip scanned). Assuming a deposit 14 mm (14,000 μm) wide, a field diameter of 250 μm, a throughput of 10 liters/minute ($0.01M^3$), and a total operating period of 24 hours (1440 minutes), this volume may be calculated in the following way:

$$\text{Volume } (M^3) = \frac{250 \ \mu m}{14,000 \ \mu m} \times 0.01 M^3/min \times 1440 \ min = 0.257 M^3$$

Even this integrated estimate is flawed, though, because some deposition occurs outside the 14-mm strip, and particle deposition within its boundaries often varies systematically, so that even adjacent longitudinal traverses may differ significantly (21). The use of at least 100 random fields distributed both parallel and perpendicular to the longitudinal axis of each tape has been proposed to eliminate this bias.

Alternatively, transverse scans, usually 12, at regular intervals may be averaged to provide a mean value for particle levels within a diurnal cycle. Although deposition patterns tend to be broader than orifice dimensions would predict, the resulting errors are small in calculating daily means. Especially for extremely small particles, however, time discrimination may be possible only to within 1 to 3 hours. Levels may be recorded directly as "particles retained per M^3 of air sampled" or normalized, when appropriate, as a percentage of the mean value over a relevant period.

Because intake efficiency varies with particle size and ambient air speed, these factors do affect the accuracy of recovery data as indicators of actual particle prevalence. Under most weather conditions, though, the Burkard trap with its standard 14 × 2 mm orifice (13 mph intake speed) is believed to collect recognizable fungus particles with efficiencies exceeding 50%. A smaller interchangeable orifice may be substituted to facilitate collection of extremely minute particles, but its capacity to recover familiar fungus allergens has been less well studied.

Several alternative collectors, more or less resembling the Hirst trap, have been developed to exploit particle impingement on greased solid surfaces. Wind-oriented traps (Lanzoni, Bologna, Italy; G.R. Electric Manufacturing Company, Manhattan, KN) are suitable for outdoor application. In addition, several units with fixed orifices are suitable for collection in relative still air (24,27,49). Recoveries by most of these samplers provide time discrimination; furthermore, several provide discrete bands of deposition at predetermined intervals (24,27). The latter units have special value for investigations relating fungus particle prevalence precisely to environmental events.

Volumetric Viable Collectors

Although gravitational methods for culturing fungi from air still attract proponents (22), the deficiencies of this approach have been generally acknowledged (8,16,38,41,44). There are numerous volumetric alternatives to the open Petri plate; most are either slit samplers or sieve impingers. Certain of these samplers use a moving collection surface so that the time of each recovery can be finely resolved, and virtually all ensure deposition directly on growth media, a feature that tends to conserve viable units. Although a few devices may be autoclaved, essentially all can be sterilized adequately with volatile microbicides such as propylene oxide. Unfortunately, none of the widely used viable collectors is easily wind-oriented for use in rapidly moving outdoor air. In addition to units described below, culture samples may be derived using membrane filters and fluid-filled impingers.

Slit Samplers.   The impingement of aerosol units on a solid surface positioned a measured distance behind a narrow intake slit has been described. If the particles are potentially viable and the surface is a suitable sterile medium, growth points may be used to estimate prevalence as colony-forming units (CFU) recovered per unit volume of processed air. The best known of the available slit samplers was described over 40 years ago for estimating short-term trends in the prevalence of airborne bacteria (4). It is currently supplied by C.F Casella and Company, London. This "Bourdillon" sampler draws air at a defined rate downward through a slit into a housing where it impinges on the surface of medium in a rotating culture dish. Turntable speed may be varied from one rotation in 30 seconds to one in 5 minutes. Models processing 30 to 700 liters per minute of air onto agar surfaces with diameters, respectively, of 10 and 15 cm are available. The former uses a single slit (with area 0.084 cm^2); the latter uses four slits, (each 0.445 cm^2). Both the throughput rate and the distance from slit margin to the agar surface are precisely controllable, fostering a retention capability of more than 95% for minute biogenic aerosols.

Slit samplers with similar operating characteristics but of a simpler design and using modular suction sources have been marketed in the past and are currently available (e.g., Biap sampler, Mikrobiologiska och Bioteknisk Test Produkter AB, Malmo, Sweden). In addition, devices that afford time-related collections over longer periods (using noncircular, medium-coated surfaces) have been used (3,12).

Sieve Impingers. The interposition of multiply perforated plates in a flow channel promotes acceleration and impingement of viable particles so that the resulting colonies are optimally separated. If plates with holes of diminishing size are arranged serially, progressively smaller aerosol fractions may be raised to critical velocities. The development of a stacked sieve impinger by Andersen applied this principle effectively to the collection and sizing of viable airborne agents (2). Andersen samplers (Andersen 2000 Inc., Atlanta) consisting of eight, six, or two sieve plates ("stages") have been employed widely (Fig. 4–8).

Culture dishes, poured to a specific depth, may be positioned beneath each sieve and intercept all approaching particles with stop distances too long to permit escape to a downstream stage. Although size restriction is not absolute (14), particles of similar aerodynamic dimensions tend to produce growth on the same plate(s). Because such separation has limited value for studies of fungus allergens simpler arrangements, such as a plate beneath stage 6 only (or beneath stages 3 and 6), as well as the two-stage version, have been tested. If sampling times are adjusted suitably, fungus CFU levels calculated from a single sixth-stage

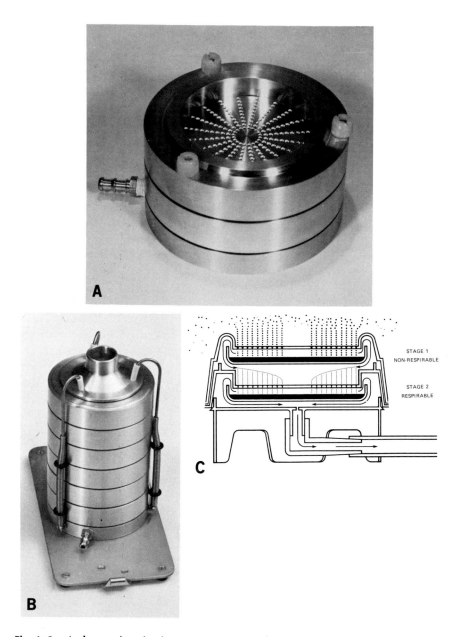

STAGE 1
NON-RESPIRABLE

STAGE 2
RESPIRABLE

Fig. 4–8. Andersen sieve impingers. *A.* Two-stage device. *B.* Six-stage device. *C.* Air and particle behavior during operation of the two-stage device. (Courtesy of Andersen Samplers, Inc., Atlanta.)

plate and from a full six-plate array have appeared to be similar (47). Comparisons of the two-stage device with other collectors are generally unavailable. In limited trials, however, a previously available disposable two-stage Andersen sampler gave consistently lower bacterial recoveries than an adjacent eight-stage unit (11).

Andersen samplers use specially designed flat-bottomed culture dishes that require 28 ml of agar medium to form correctly positioned collecting surfaces. Throughputs of 1 ft³ (28.3 liters) per minute are maintained, which usually produce shallow depressions in the agar due to "jet" flow through each perforation. Following exposure, sample dishes are covered and incubated appropriately. Counts of CFU are made visually and, for types with numerous colonies per plate, corrections for the probability of multiple impactions at single growth points are read from a "positive hole" nomogram (2). Retention of particles by multistage Andersen samplers is quite high (usually > 95%), unless overloading occurs due to excessive aerosol density or unduly long exposure periods. Losses, especially on the stage 1 hardware, have been recognized with plates at all stages (30). Like other samplers with fixed orifices, the intake efficiency of these devices is compromised seriously during exposures in rapidly moving (e.g., outdoor) air. Efforts to wind-orient multistage sieve impingers on large vanes with intakes in (or approaching) the vertical plane have improved results substantially; the feasibility of similarly adapting the two-stage unit remains to be tested (45).

The recent introduction of the Microban sampler (Ross Industries, Midland, Virginia) has provided a single-plate sieve device suitable for viable collections in relatively still air (17). This unit draws 0.7 ft³ (20 liters)/minute through a sieve plate bearing 325 holes with impingement onto a 100-mm diameter plastic Petri plate containing 15 ml of agar medium; a pump and flow adjustment are integral components. Relative to the two-stage Andersen collector, the Microban sampler offers economy of initial cost of materials as well as a more compact and portable form. Adaptability of the latter device to differing flow situations, though, may be more limited; furthermore, its collection efficiency for common fungus particles remains uncertain, although bacterial recoveries by this unit and by a six-stage Andersen sampler were similar in limited comparisons (23).

Additional Suction Collectors

Because the concerns of fungus allergy naturally focus on more common air spora, high throughput devices generally receive little attention. Present initiatives in the immunochemical assay of aeroallergens will certainly involve high-volume samplers with filters providing almost absolute retention of *aspirated* micronic particles (1). Problems incident to the fixed intake orientation of these devices remain unsolved, however, and predispose sampling to losses, especially of larger aerosols.

A novel approach to recovering spores from large air volumes has been provided by Schwarzbach's jet impinger (42). With adequate suction applied, this device aspirates over 450 liters per minute. The sampled air is accelerated by an aerodynamically adapted jet and forced against the open orifice of a tube containing a still air column. This tube, in turn, opens downward into a sealed settling chamber in which particles may sediment onto moist plant materials or other culture substrates. Retention of large (>20 μm) particles is reported to be high, although intake efficiencies and small aerosol handling await estimation. Both the Scnwarzbach jet sampler and the collection principles it exploits may have alternative uses in the analysis of air spora.

There are other samplers with potential application to problems of mould allergy; mention will be made here only of the Biotest (RCS) centrifugal sampler (Folex-Biotest-Schleussner, Moonachie, New Jersey). This DC-powered handheld device processes 1.4 ft³ (39.6 liters) of air per minute, with intake and depositon of particles on agar controlled by a centrifugal impeller. Unlike its ponderous ancestor, the Wells air centrifuge, the RCS sampler is lightweight, potentially useful as a sampling "probe," and may admit a measure of wind orientation. During limited trials in relatively still air, bacterial recoveries by the RCS sampler have been reported to be comparable to those of commercial slit samplers, such as the Bourdillon and Microban (17).

SAMPLING AIRBORNE FUNGI: CURRENT STATUS AND PROSPECTS

The foregoing survey must emphasize just how large the variety of usable methods has become. This diversity reflects both the many facets of mould allergy deserving study and the inadequacy of any one or two devices to answer all of these investigative needs. Furthermore, the cost of several samplers ideally suited for specific applications is now so high as to preclude their use effectively. Accordingly, three major objectives for the future seem fundamental: the formulation by clinicians of finely focused questions concerning the ecology of mould exposure; the development of increasingly cost-effective collection methods; and an improved awareness by clinicians and others of the potentials and limitations of each approach considered.

In surveying current methods, the division between outdoor and indoor sites is both natural and useful pragmatically. In outdoor air data from spore traps (such as the Burkard trap) describe the broadest range of particles and provide a time dimension in which the impact of events on prevalence may be discerned. Impactors (e.g., the intermittent rotorod) are properly favored when averages for particles above 5 to 8 μm will suffice, giving recoveries roughly equivalent to those from the Burkard trap for large aerosols (45); however, suction samplers should remain the choice for studies of small spores. Wherever possible, direct microscopic identification of particles is to be preferred, with culture reserved for types without distinctive airborne units. Increasing efforts to identify preferred media and culture conditions, with attention to selective growth inhibitors, are strongly indicated at present as is the development of a viable collector adaptable to wind orientation (6). In addition, there is an increasingly evident need to secure extensive data at ground level, relating fungus prevalence to natural plant communities, land use, and human activities.

Indoor sampling efforts confront a different array of problems, not the least of which is the distinction between interior and outdoor sources for the recoveries obtained (39,44). Because small-spored taxa (such as *Penicillium* and *Apergillus* sp.) often predominate indoors, viable sampling methods assume overriding importance. Furthermore, because small particles are least easily recovered by traditional open plate methods, an economical volumetric approach, suited to culture, is needed. Although the two-stage Andersen collector and Microban sampler offer the aerometric potential required, their cost leaves little hope that they will be widely used in patients' homes. Solutions to this problem may come through adaptations of open filters or liquid impingers operated through critical orifices by small air movers. Miniature pumps also hold the promise that truly volumetric personal monitors may be evolved to individualize assessments of exposure risk further.

REFERENCES

1. Agarwal, M.K., et al.: An immunochemical method to measure atmospheric allergens. J. Allergy Clin. Immunol., **68**:194, 1981.
2. Andersen, A.A.: New sampler for the collection, sizing and enumeration of viable airborne particles. J. Bacteriol., **76**:471, 1958.
3. Andersen, A.A., and Andersen, M.R.: A monitor for airborne bacteria. Appl. Microbiol., **10**:181, 1962.
4. Bourdillon, R.B., Lidwell, O.M., and Thomas, J.C.: A slit sampler for collecting and counting airborne bacteria. J. Hyg. (Camb.), **41**:197, 1941.
5. Burge, H.P., et al.: Comparative recoveries of airborne spores by viable and nonviable modes of volumetric collection. Mycopathol., **61**:27, 1977.
6. Burge, H.P., Solomon, W.R., and Boise, J.R.: Comparative merits of eight popular media in aerometric studies of fungi. J. Allergy Clin. Immunol., **60**:199, 1977.
7. Cadham, F.T.: Asthma due to grain rusts. JAMA, **83**:27, 1924.
8. Cole, W.R., and Bernard, H.R.: Quantitative air sampling, a contrast with the settling-plate method for the study of bacterial air contamination in operating rooms. Surgery, **51**:658, 1962.
9. Cour, P.: Nouvelles techniques de détection des flux et des retombées polliniques: étude de la sedimentation des pollens et des spores à la surface du sol. Pollen et Spores **16**:1,103, 1974.
10. Cryst, S., Gurney, C.W., and Hansen, W.: A method for determining aeroallergen concentrations with the molecular filter membrane. J. Lab. Clin. Med., **46**:471, 1955.
11. Curtis, S.E., Balsbaugh, R.K., and Drummond, J.G.: Comparison of Andersen eight-stage and two-stage viable air samplers. Appl. Environ. Microbiol., **35**:208, 1978.
12. Decker, H.M., et al.: Design and evaluation of a slit-incubator sampler. Appl. Microbiol., **6**:398, 1958.
13. Durham, O.C.: The volumetric incidence of airborne allergens. IV. A proposed standard method of gravity sampling, counting and volumetric interpolation of results. J. Allergy, **17**:79, 1946.

14. Flesch, J.P., Norris, C.H., and Nugent, A.E., Jr.: Calibrating particulate air samplers with monodisperse aerosols: Application to the Andersen cascade impactor. Am. Ind. Hyg. Assoc. J., **28**:507, 1967.
15. Fuchs, N.A.: The Mechanics of Aerosols. London, Pergamon Press, 1964.
16. Gregory, P.H.: The Microbiology of the Atmosphere. 2nd Ed. New York, John Wiley and Sons, 1973.
17. Groschel, D.H.M.: Air sampling in hospitals. Ann. N.Y. Acad. Sci., **353**:230, 1980.
18. Harrington, J.B., Gill, G.C., and Warr, B.R.: High efficiency pollen samplers for use in clinical allergy. J. Allergy, **30**:357,. 1959.
19. Hirst, J.M.: An automatic volumetric spore trap. Ann. Appl. Biol., **39**:257, 1952.
20. Kailin, E.W.: Variations in ragweed pollen in a metropolitan area. Med. Ann. D.C., **33**:1, 1964.
21. Kapyla, M., and Penttinen, A.: An evaluation of the microscopical counting methods of the tape in Hirst-Burkard pollen and spore trap. Grana, **20**:131, 1981.
22. Kethley, T.W., and Cown, W.B.: In defense of settling plates. Hosp. Man., **103**:84, 1967.
23. Kotula, A.W., et al.: Comparison of a single and a multiple stage sieve sampler for airborne microorganisms. Unpublished. (Courtesy of Ross Industries, Midland, VA.)
24. Kramer, C.L., and Pady, S.M.: A new 24-hour spore sampler. Phytopathol., **56**:517, 1966.
25. Kramer, C.L., and Pady, S.M.: Viability of airborne spores. Phytopathol., **60**:448, 1968.
26. Leuschner, R.M., and Boehm, G.: Investigations with the individual pollen collector and Burkard trap with reference to hay fever patients. Clin. Allergy, **9**:175, 1979.
27. Marx, H.P., Spiegelman, J., and Blumstein, G.I.: An improved volumetric impinger for pollen counting. J. Allergy, **30**:89, 1959.
28. May, K.R.: The cascade impactor: An instrument for sampling coarse aerosols. J. Sci. Instr., **22**:187, 1945.
29. May, K.R.: A cascade impactor with moving slides. Arch. Ind. Hlth., **13**:481, 1956.
30. May, K.R.: Calibration of a modified Andersen bacterial sampler. Appl. Microbiol., **12**:37, 1964.
31. May, K.R.: Physical aspects of sampling airborne microbes. *In* Airborne Microbes. Edited by P.H. Gregory and J.L. Monteith. Cambridge, Cambridge University Press, 1967.
32. May, K.R. and Druett, H.A.: The pre-impinger, a selective aerosol sampler. Br. J. Ind. Med., **10**:142, 1953.
33. May, K.R., and Harper, G.J.: The efficiency of various liquid impinger samplers in bacterial aerosols. Br. J. Ind. Med., **14**:287, 1957.
34. McDonald, J.L., and Solomon, W.R.: Effect of outdoor activity on aeroallergens in the human microenvironment (abstract). J. Allergy Clin. Immunol., **55**:89, 1975.
35. Ogden, E.C., and Raynor, G.S.: Field evaluation of ragweed pollen samplers. J. Allergy, **31**:307, 1960.
36. Ogden, E.C., and Raynor, G.S.: A new sampler for airborne pollen: The Rotoslide. J. Allergy **40**:1, 1967.
37. Raynor, G.S.: An isokinetic sampler for use on light aircraft. Atmos. Environ., **6**:191, 1972.
38. Raynor, G.S.: Sampling techniques in aerobiology. *In* Aerobiology. The Ecological Systems Approach. Edited by R.L. Edmonds. Stroudsburg, Dowden, Hutchinson & Ross, 1979.
39. Richards, M.: Atmospheric mold spores in and out of doors. J. Allergy, **25**:429, 1954.
40. Ryan, K.J., and Mihalyi, S.F.: Evaluation of a simple device for bacteriologic sampling of respirator-generated aerosols. J. Clin. Microbiol., **5**:178, 1977.
41. Sayer, W.J., Dudley, B.S., and Jamshid, G.: Estimation of airborne fungal flora by the Andersen sampler versus the gravity settling culture plate. I. Isolation frequency and numbers of colonies. J. Allergy, **44**:214, 1969.
42. Schwarzbach, E.: A high throughput jet trap for collecting mildew spores on living leaves. Phytopathol. Z., **94**:165, 1979.
43. Solomon, W.R. Unpublished oservations.
44. Solomon, W.R.: Assessing fungus prevalence in domestic interiors. J. Allergy Clin. Immunol., **56**:235, 1975.
45. Solomon, W.R., et al.: Comparative particle recoveries by the retracting rotorod, Ro-

toslide and Burkard spore trap sampling in a compact array. Int. J. Biometeorol., **24**:107, 1980.

46. Solomon, W.R., Burge, H.A., and Boise, J.R.: Performance of adhesives for rotating arm impactors. J. Allergy Clin. Immunol., **65**:467, 1980.
47. Solomon, W.R., and Gilliam, J.A.: A modified application of the Andersen sampler to the study of airborne fungus particles. J. Allergy, **45**:1, 1970.
48. Solomon, W.R., Stohrer, A.W., and Gilliam, J.A.: The "fly-shield" rotobar: A simplified impaction sampler with motion-regulated shielding. J. Allergy, **41**:290, 1968.
49. Stedman, O.J.: A seven-day volumetric spore trap for use within buildings. Mycopathol., **66**:37, 1978.
50. Swinney, B., Sr., Swinney, B., Jr., and Hicks, R.: A simple improved pollen sampler. Ann. Allergy, **26**:605, 1968.
51. Watson, H.H.: Errors due to anisokinetic sampling of aerosols. Am. Ind. Hyg. Assoc. Q., **15**:21, 1954.

Chapter 5

THE CONIDIA

Garry T. Cole
Robert A. Samson

Conidial fungi, which comprise the form-subdivision Deuteromyco-tina, are generally the most important group of air-disseminated fungi that cause respiratory allergic diseases in humans (56,100). Investigations of fungal air spora, irrespective of the type of spore trap employed or the geographic region examined, have demonstrated that conidium-forming fungi usually predominate over other fungal groups. Exceptions occur in those areas that are particularly conducive to growth of basi-diomycetous fungi, such as New Orleans (116–118) and Auckland, New Zealand (S.M. Hasnain, personal communication), where at certain periods of the year basidiospores are the predominant airborne fungal propagules. Lacey (85), however, drawing from the results of about 200 air spora studies in different parts of the world, has concluded that *Cladosporium, Alternaria, Penicillium,* and *Aspergillus,* listed in decreasing frequency of occurrence, are most consistently associated with the highest mean percentages of total fungal spore catches. It is not surprising that these same four genera represent the most common allergenic moulds, based on skin reactivity studies (100). Approximately 85% of patients allergic to moulds will react to one or more of these allergens.

The kinds of substrates within or on which these moulds can grow and sporulate is necessarily of concern to anyone investigating or suffering from mould allergies. If provided reasonable protection and adequate moisture, many microfungi can grow and sporulate on nutritionally marginal substrates (128). In addition to daily variations in meteorologic conditions and seasonal changes, both of which have enormous impact on the concentration of air spora (58,103), the amount and type of vegetation of a region or microenvironment may be important factors in determining the composition of the airborne fungal population (85). Of course, biotic factors such as human activity, which can also have significant effects on atmospheric spore concentrations, must not be overlooked. For example, mowing and haymaking often result in dramatic increases in numbers of airborne spores, especially conidia of *Cladosporium* and *Epicoccum* (58). Dense clouds of these fungal propagules may be dispersed by the wind and subsequently elicit allergic responses from sensitized individuals who are far from the source.

In most cases the composition of air spora is based on straightforward classification and on counting procedures of trapped spores using one of several types of instruments available (58, 147). An alternative to enumeration of air spora is a consideration of volume contributions of different taxa to the total volume of the spore catch. Under these conditions, the predominating conidial fungi are commonly *Alternaria, Curvularia, Helminthosporium,* and *Epicoccum,* but are often overshadowed by larger airborne ascospores and basidiospores. Lacey (85) has emphasized the relevance of such a comparison to investigations of the allergenic potential of air spora. He pointed out that conidia and other fungal spores associated with an immediate-type allergic response, as demonstrated by atopic inidividuals, are usually larger than 5 μm, while those associated with delayed-type allergic response are considerably smaller, ensuring deeper penetration into the lungs (Table 5–1). Also of significance is whether or not the conidia enter the respiratory tract as individual propagules or as aggregates (e.g., spore droplets or chains). For example, clusters of small conidia of *Aspergillus fumigatus* (individual conidia measure 2 to 3.5 μm in diameter) may be deposited in the upper respiratory tract.

Gregory (58) has compared the human respiratory system to an air sampler in which different sized particles are taken up nonselectively and then sorted by various deposition mechanisms acting in succession. The site at which spores are deposited depends on the anatomy of the respiratory tract and on the terminal velocity of the particles (i.e., on their aerodynamic size). The size, density, and surface topography of airborne spores are, therefore, important considerations in studies of the distribution of aeroallergens within the respiratory tract (Fig. 5–1). Most large spores (>10 μm) are entrapped in the mucous layer of the nasal labyrinth during normal nose breathing and are then swept proximally toward the nasopharynx by the continuously moving cilia (Fig. 5–1A). On the other hand, fungal spores in the range of 1 to 5 μm often reach the lower regions of the respiratory tract. Much higher percentages of larger spores are deposited in the trachea and bronchi when the nasal labyrinth is bypassed during mouth breathing (Fig. 5–1A and B).

Of uncertain significance is the nature of chemical interactions between the fungal spore envelope and lung surfactant, as well as other

TABLE 5–1. Representative Conidial Fungi Reported as Causative Agents of Human Allergies

Reference No.*	Taxon	Conidial Size (L × W), μm	Conidial Shape	Conidial Formation and Arrangement in Culture†	Natural Substrates
56, 77	Alternaria alternata‡	18–83 × 7–18	Obpyriform with cylindric beak	From localized regions of fertile cells (porogenous); in chains	Cosmopolitan, decaying plant material, foodstuffs, soil, textiles
2	Arthrinium sphaerospermum	7–9	Globose to lenticular	Holoblastic from basauxic conidiophores; singly	Grasses (Phleum, Arundo)
77	Aspergillus candidus‡	2.5–4	Globose to subglobose	From phialides on distinct fertile vesicle (ampulla); in chains or slimy heads in fresh isolates	Stored cereals, grains and grain products (mouldy grain); 2° decomposer of vegetation; warm soils from tropical and subtropical areas
85, 115	A. clavatus‡	3–4.5 × 2.5–4.5	Ellipsoid	From phialides; several divergent columns	Soils, animal dung
71, 77, 115	A. flavus‡,§	3–6	Globose to subglobose	From phialides; in columns (false chains)‖	Mouldy corn, oil seeds, grains, groundnuts, forage and decaying vegetation, dairy products, foodstuffs, warm soils
71, 77, 115	A. fumigatus§	2–3.5	Globose to subglobose	From phialides; in columns	Common in composts and other organic material undergoing decomposition at high temperatures (40° C), warm and temperate soils, cereals
77	A. nidulans	2.5–4	Globose	From phialides; in short columns	Primarily soil (temperate and subtropical), slowly decomposing vegetation
71, 77, 115	A. niger‡	3.5–5	Globose to subglobose	From phialides; in loose columns at maturity	Worldwide on a great variety of substrates, including grains, forage products, spoiled fruits and vegetables, dairy products, textiles, and warm soils

77	A. ochraceus§	2.5–3	Globose to subglobose	From phialides; in compact columns	Grains, soil, salted foodstuffs (not active decomposers of vegetation)
110	A. penicilloides	3–3.5 × 4–5	Ellipsoid, barrel-shaped	From phialides; in columns	Foodstuffs, xerophilic habitats (house dust)
2	A. glaucus (= A. repens, teleomorph Eurotium herbariorum)	5–6.5	Globose to subglobose	From phialides; in columns	Decaying plant material stored at very low moisture levels; seeds, grains, food products containing high concentrations of sugar; meat; wool
2	A. terreus‡,§	1.8–2.4	Globose to slightly ellipsoid	From phialides; in compact columns	Primarily from warm soil, decaying vegetation; also stored grain, straw, cotton
77	A. versicolor‡	2–3.5	Globose	From phialides; in loose columns	Stored barley, hay, cotton, cheese; cured meats; common in a variety of soils
56	Aureobasidium pullulans§	1° conidia 9–11 × 4–5.5; 2° conidia 4–6 × 2–3; chlamydospores (conidia?) 7.5–16 × 3.5–7	Ellipsoid to ovoid 1° conidia; ellipsoid 2° conidia; cylindric to spheroid chlamydospores	1° conidia from phialides, singly and in clusters; 2° conidia by budding of 1° conidia, in slimy colonies; chlamydospores from mycelia by thallic development; in chains	Ubiquitous and omnivorous; decaying pears and oranges in storage, paint, wood and paper, soil
56	Botrytis aclada (= B. allii)	7–11 × 5–6	Narrowly ellipsoid	Synchronous holoblastic development from fertile branches; botryose	Onions and shallots in field and storage
56	B. cinerea	8–14 × 6–9; microconidia 2.5 × 3	Obovoid; microconidia globose	As B. aclada microconidia from phialides; in droplets	Facultative parasite of wide range of plants (e.g., "grey mould" of grapes, strawberries, cabbage, lettuce); soil, stored fruits, vegetables
77	Cladosporium cladosporioides	3–11 × 2–5	Ellipsoid to limoniform	Holoblastic development in acropetal succession; in true chains	Cosmopolitan; soil, textiles, foodstuff, stored crops

TABLE 5–1. *Continued*

Reference No.*	Taxon	Conidial Size (L × W), μm	Conidial Shape	Conidial Formation and Arrangement in Culture‡	Natural Substrates
56	C. herbarum	5–23 × 3–8	Ellipsoid to cylindric	As C. cladosporioides	Abundant in temperate regions on dead herbaceous and woody plants (straw), soil, foodstuffs, paint, textiles
56	C. macrocarpum	9–28 × 5–13	Ellipsoid	As C. cladosporioides, except chains shorter	As C. herbarum
77	C. sphaerospermum	3–4.5	Globose to subglobose	As C. cladosporioides	2° invader of many different plants, soil, foodstuffs, paint, textiles
115	Cryptostroma corticale	4–6.5 × 3.5–4	Ellipsoid to obovoid	From phialides; in chains	Bark of maple (North America) and sycamore (Acer pseudoplatanus in Great Britain), stored logs (sooty bark disease)
115	Drechslera sorokiniana	40–120 × 17–28	Fusiform to broadly ellipsoid	From localized regions of fertile cells (porogenous); singly	Cosmopolitan; on grasses, cereals, including barley, oats, rye, and wheat, decaying foodstuff
56	Epicoccum purpurascens	15–25	Globose to pyriform	Holoblastic, formed terminally and laterally from fertile hyphae; singly and in clusters	2° decomposer of plants, soil, caryopsis (corn, barley, oats, wheat), beans, paper, textiles
2	Fulvia fulva (= Cladosporium fulvum)	12–47 × 4–10	Ellipsoid	Holoblastic development in acropetal succession; in true chains	Leaf mould on tomatoes, especially on plants grown in greenhouses
56, 117	Fusarium solani‡	Macroconidia, 27–52 × 4.4–6.8; microconidia, 8–16 × 2–4	Macroconidia fusiform, cylindric, slightly curved; microconidia ovoid to oblong	From phialides; singly or in droplets	Cosmopolitan; soils, wide range of plants

117	Geotrichum candidum§	6–12 × 3–6	Cylindric barrel-shaped, or ellipsoid	Fragmentation of fertile hyphae; in chains	Cosmopolitan; soil, water, cereals, rice, grapes, citrus fruits, bananas, tomatoes, cucumber, bread, milk products, paper, textiles
77	Penicillium brevicompactum	2.5–3.5 × 2–2.5	Globose to subglobose	From phialides; disordered chains, resembles Aspergillus	Soil, decaying vegetation, foods, cereals, textiles, paints
77	P. citrinum (= P. steckii)‡	2.5–3	Globose to subglobose	From phialides; in columns	Very common, soil, food
77	P. decumbens	2.5–4 × 2–3	Ellipsoid to pyriform	From phialides; short loose columns	Soil
77	P. expansum‡	3–3.5 × 2.5–3	Subglobose to ellipsoid	From phialides; densely packed, irregular chains	Decaying pomiferous fruit
85	P. frequentans	3–3.5	Globose to subglobose	From phialides; long columns	Extremely common; soil, composts, animal feces, cereals, foodstuffs, paper and paper pulp, paint
71, 115	P. herquei	3.5–5 × 3.5	Ellipsoid to apiculate	From phialides; columns	Soil
77	P. implicatum	2.5–3 × 2–2.5	Ellipsoid	From phialides; loose columns	Soil
71, 115	P. italicum	4–5 × 2.5–3.5	Cylindric to ellipsoid or subglobose	From phialides; long disordered chains	Primarily citrus fruits; also soil and decaying vegetation
77	P. oxalicum	3.5–7 × 2.5–4	Ellipsoid	From phialides; long columns	Soil, decaying vegetation
85	P. roquefortii (= P. casei)‡	4–6	Globose to subglobose	From phialides; loose columns	Stored foods at cool temperatures, cheese, rye bread
71, 115	P. simplicissimum	2.5–3.5 × 2–3.5	Ellipsoid	From phialides; disordered chains	Decaying vegetation, widely distributed in soils, foodstuffs
77	P. verrucosum var. cyclopium (= P. aurantiogriseum = P. cyclopium)‡	3–4	Globose to subglobose	From phialides; false chains	Very common, drying crops, food
77	P. verrucosum var. verrucosum (= P. viridicatum)	3–4	Globose to subglobose	From phialides; irregular chains	Cereal grains and derived products

TABLE 5–1. *Continued*

Reference No.*	Taxon	Conidial Size (L × W), μm	Conidial Shape	Conidial Formation and Arrangement in Culture†	Natural Substrates
77	*P. waksmanii*	2.5–3.5	Globose	From phialides; short irregular chains	Soil
18	*Trichophyton mentagrophytes*§	Macroconidia, 6–8 × 20–50; microconidia 2.5–4 diam., 2–3 × 3–4; arthroconidia 4–5 × 2–3	Clavate macroconidia; spherical to clavate microconidia; cylindric arthroconidia	Macro- and microconidia by holothallic development, singly; arthroconidia by fragmentation of fertile hyphae, chains	Soil, keratin
56	*Trichothecium roseum*‡	12–23 × 8–10	Ellipsoid to pyriform	Retrogressive, blastic development; chains	Cosmopolitan; decaying vegetation, soil, seeds of corn, flour products
56, 85	*Verticillium lecanii*	2.3–10 × 1–2.6	Cylindric to ellipsoid	From phialides; in droplets	Decaying vegetation (straw), soil, entomogenous on various arthropods
77	*Wallemia sebi*	2.5–3.5	Globose to subglobose	From basauxic conidiophore; in chains	Cosmopolitan; foodstuffs such as jams, dates, bread, cakes, salted meats, dairy products, fruit; soil, hay, textiles

*Reference sources for representative allergenic conidial fungi are given here next to each taxon.

†For descriptions and illustrations of the methods of conidium formation, see Cole and Samson (39).

‡Produces toxins that may be associated with disease in humans and in other animals (*Alternaria alternata*: tenuazonic acid and other toxic metabolites (93); *Aspergillus candidus* and *A. clavatus*: patulin (93); *A. flavus*: aflatoxin, possibly hepatocarcinogenic (93); *A. ochraceus*: penicillic acid, ochratoxin A, xanthomegnin, and viomellein, all reported as both hepatotoxic and nephrotoxic (93); *A. terreus*: patulin and citrinin (93); *A. versicolor*: cyclopiazonic acid, sterigmatocystin (93); *Fusarium solani*: trichothecenes (93); *Penicillium verrucosum* var. *cyclopium*: penicillic acid, cyclopiazonic acid, penetrem A, B (93); *P. citrinum*: citrinin (93); *P. expansum*: penicillic acid and patulin (93,96); *P. roquefortii*: PR toxin, roquefortine (93,96); *Trichothecium roseum*: trichothecenes (93). See also Cole and Cox (41).

§Pathogenic in humans: *Aspergillus flavus*, *A. fumigatus*, *A. nidulans* and *A. terreus* associated with aspergillosis of the lungs and/or disseminated aspergillosis (115); *A. niger* frequently encountered in otomycosis (115); *Aureobasidium pullulans* associated with phaeomycotic cutaneous colonization (115); *Geotrichum candidum* causative agent of pulmonary, bronchial, oral, gastrointestinal, and cutaneous geotrichosis (115); *Trichophyton mentagrophytes* associated with various dermatophytoses (115).

‖"False" chain is a basipetal chain of conidia in which cells are held together by wall connectives (e.g., *Aspergillus*, *Penicillium*; see Fig. 5–6) but are not united cytoplasmically (c.f. "true" chains; see Cole and Samson (39)).

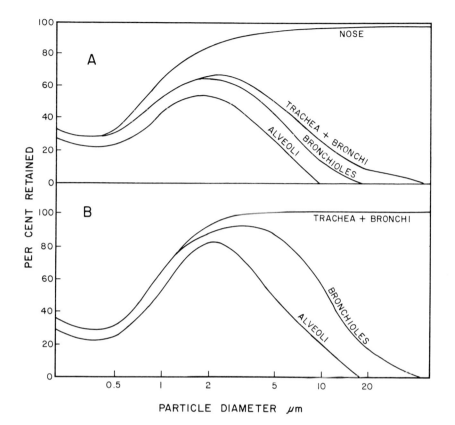

Fig. 5–1. Retention of inhaled particles in the respiratory tract. Relationship between particle size and site of deposition is shown for normal nose breathing *(A)* and for mouth breathing *(B)*. (Modified from Lidwell (88) and Gregory (58).)

secretory products associated with the respiratory mucosa (83). The hydrophobic surface layer of rodlet fascicles (see Fig. 5–9), which characterizes most conidia and certain sporangiospores (35,43,59,60,66,68,69), provides an advantage for air dispersal (16) and perhaps some protection against host defense mechanisms (76). In the presence of commercial lung surfactant (such as phosphatidylcholine dipalmitoyl), however, cell components that are presumed to be of wall origin are released through the intact rodlet layer (see Fig. 5–13 and Table 5–3). Results of examinations of the chemical composition and immunogenicity of fungal cell walls (55,63,64,120,141) suggest that a large reservoir of potentially active antigens resides in the wall and may be released during initial contact and subsequent fungus-host interaction. Current research on the nature of pulmonary surfactant and mucous secretions using organ and tissue cultures (111), tracheal explants (86), lung-originated cell monolayer cultures (15), and cultures of purified cell populations [i.e., alveolar epithelial type II cells (84)] is providing the technology for development of models to investigate initial biochemical events associated with the impaction and sedimentation of fungal propagules on the mucosa of the host respiratory tract.

The focus of the following discussion is on the conidium, the asexual, nonmotile, usually deciduous propagule produced by deuteromycetous fungi (39), and its significance in respiratory allergies. Our approach is that of mycologists rather than allergists, although we will direct our attention to biologic features of the conidium that have established or implicated associations with allergic diseases. These include ecologic aspects of conidial fungi as well as size, topography, surface ultrastructure, and wall chemistry of conidia.

NATURAL SUBSTRATES

Gregory (58) has stated that "outdoor airborne allergens appear to originate from vegetation above ground, but not significantly from the soil . . . . To a large extent, therefore, their origin is an agricultural problem aggravated by our needs to cultivate plants in pure stands" and to store them subsequently for long periods (119). Nevertheless, the significance of soil as a natural ecologic habitat of conidial fungi should not be underemphasized (13). It is a reservoir of microorganisms, some of which grow and sporulate within the soil and play important roles in breakdown of organic debris, while others are transient, primarily colonizing and invading available organic surface substrates. For example, members of the *Aspergillus nidulans* group are considered to be primarily of soil origin, occurring both in temperate and subtropical soils. On the other hand, members of the *A. clavatus* group commonly grow on dung of various animals, and their presence in the soil may be the result of association with such substrates (110). The sources of indoor air spora are air conditioning systems, domestic humidifiers, pets, and other substrates (25,129). The soil of potted ornamental plants is also an important reservoir of pathogenic and allergenic species, such as *A. fumigatus* (130,131,132,140).

Isolation of certain fungi from the soil that have evolved specific substrate relationships, such as keratinophilic *Trichophyton* spp., necessitates the use of "baiting" techniques and selective media (80). Encountering facultative parasites of higher plants in the soil is likely to be due to release of rain-washed spores from the host. Because they compete with indigenous soil saprobes, it would be expected that these parasitic forms could not survive in this essentially foreign or casual habitat (13). *Penicillium italicum* is mainly encountered as the causative agent of blue mould rot of citrus fruits, and is only occasionally isolated from soil. Similarly, *P. expansum* is commonly isolated from decaying pomiferous fruits and its detection in the soil is rare. Because *Wallemia sebi* is osmophilic, this fungus is most frequently isolated from substrates with a high sugar or salt content, such as jams and salted meats. These fungi have become specially adapted for growth on epigeous substrates, and in such cases the soil constitutes a "primeval habitat" (106).

On the basis of the natural substrates cited for most of the 44 representative allergenic conidial fungi in Table 5–1, the primary origin of airborne fungal allergens does seem to be organic aboveground substrates. It is interesting to note that most of these allergenic species also represent the common fungal contaminants of foodstuffs (119). Agricultural crops are common substrates for fungal growth, and the numbers of airborne conidia may be significantly increased by human activities, such as harvesting (85). *Aspergillus candidus* is economically important as a causative agent of mouldy grain. It is a successful competitor in mixed populations of fungi on wheat stored at 75 to 85% relative humidity and is able to invade and sporulate on stored seeds extensively in less than 30 to 35 days (99). Subsequent handling and packaging of such contaminated grain releases high levels of potential allergens into the air. *A. flavus* is also commonly encountered on stored crops such as grain, oil, and seeds but requires slightly higher minimal moisture levels than *A. candidus* (78).

The survival of conidial fungi, like that of most other organisms, depends on the availability of water, a factor that also influences the selection of natural substrates on which these allergenic moulds occur. The concept of water activity (a_w) as a factor influencing saprophytic fungal growth was introduced to microbiologists by Scott (123) and has been reviewed by Pitt (105,107). Water activity is numerically equal to the equilibrium relative humidity *(ERH)* expressed as a decimal, so that

$$a_w \text{ (organic substrate)} = ERH/100 \text{ (air)}$$

where *ERH* refers to the air. Pitt (107) has pointed out that, in many cases of biodeterioration, mould growth occurs at ambient temperatures with abundant oxygen supply, an approximately neutral pH, with adequate nutrients, and in the absence of preservatives or fungicides. Under these conditions, a_w is the predominant factor affecting fungal growth.

If the degree of tolerance of low water activity is simply expressed in terms of minimum a_w at which germination and growth can occur, an important distinction between groups of allergenic fungi can be recognized (Table 5–2). Fungi that are able to grow below $0.85a_w$ under a defined set of environmental conditions are considered to be xerophiles (105). These microbes can survive on dry organic substrates in equilibrium with an atmosphere of normal humidity while those fungi requiring minimal water activity above $0.85a_w$ are restricted to moist substrates and/or abnormally moist atmospheres. It is evident from Table 5–2 that

TABLE 5–2. Minimum Water Activities (a_w) Permitting Germination of Allergenic Conidial Fungi on Natural Substrates*

Fungus	Minimum Reported a_w	Time (days)	Temperature (° C)
Aspergillus candidus	0.75	14	25
A. flavus	0.78	ns†	33
A. fumigatus	0.82	ns	40
A. nidulans	0.78	ns	37
A. niger	0.77	ns	35
A. ochraceus	0.77	57	25
A. terreus	0.78	ns	37
A. versicolor	0.78	24	25
Botrytis cinerea	0.93	2	20
Cladosporium herbarum	0.88	7	20
Fusarium solani	0.90	56	20
Geotrichum candidum	0.90	15	20
Penicillium aurantiogriseum (= P. cyclopium)	0.81	120	25
P. brevicompactum	0.78	24	25
P. citrinum	0.80	14	25
P. expansum	0.82	22	25
P. herquei	0.88	6	25
P. implicatum	0.78	10	25
P. oxalicum	0.86	5	23
P. simplicissimum	0.86	11	25
P. viridicatum (= P. verrucosum)	0.80	21	25
Trichothecium roseum	0.90	4	20
Wallemia sebi	0.75	ns	22

*After Pitt (107).
†Not stated.

Aspergillus spp., most species of *Penicillium,* and *W. sebi* are xerophiles. These fungi are apparently capable of competing successfully with most conidial fungi over almost the entire range of water availability (107). In most cases agricultural crops, processed foods, and other organic products stored at $0.70a_w$ or below do not support fungal growth.

The incidence of moulds indoors is usually detected by the use of standard agar media, with an a_w approaching 1.00. Xerophilic fungi, however, have been shown to be abundant in house and mattress dust (89,91,92). *A. glaucus* (= *A. repens;* see Fig. 5–8), *A. penicilloides* (see Fig. 5–6), and *W. sebi* are most abundant, while the strongly xerophilic *Eurotium halophilicum,* rarely encountered in nature, has also proved to be present in house dust. The xerophilic fungal flora plays an important role in the ecology of house dust mites, such as *Dermatophagoides pteronyssinus* (Acarida: Pyroglyphidae) (90, 92), which are strongly associated with house dust allergens (24). Lustgraaf (90) has suggested that xerophilic fungi stimulate the growth of house dust mite populations. Rijckaert (114), however, has demonstrated that extracts of *A. glaucus*, *A. penicilloides*, and *Wallemia sebi* are active in human skin tests.

Temperature is also an important factor in relation to potential substrates of allergenic fungi. Both *A. fumigatus* and *A. candidus* are capable of growth at temperatures up to at least 55° C (6,48,135), an essential feature permitting their survival in composts and in other organic substrates undergoing decomposition at high temperatures. These are thermophiles, demonstrating growth optima at 40 to 45° C, but they are also thermotolerant, and can grow at temperatures as low as 10 to 15° C (107). This broad range of temperature tolerance may partly explain Barron's (13) observation that *A. fumigatus* is the most common species of *Aspergillus* isolated from temperate Canadian soils. Most conidial fungi are mesophiles, needing a minimum temperature for growth between 0 and 15° C, with an optimal temperature between 25 and 35° C and maximal temperature below 50° C (107).

A few fungi are psychrophiles, with an optimal temperature for growth between 10 and 25° C (94,98,107); some of these are recognized as allergenic forms. The latter include *Botrytis cinerea* (min., −2° C; opt., 22–25° C; max., 30–33° C), *Cladosporium herbarum* (min., −5° C; opt., 24–25° C; max., 30–32° C), *Penicillium brevicompactum* (min., −2° C; opt., 23° C; max., 30° C), and *P. expansum* (min., −2° C; opt., 23° C; max., 30° C). Such psychrotolerant fungi may be capable of growth and even sporulation on vegetation in temperate climates throughout the year, thereby providing a continual supply of airborne conidia for sensitized victims.

Considerable attention has been focused on both *A. flavus* and *A. fumigatus* due to their well-documented toxicogenic and pathogenic features (93,115). The principal substrates of *A. flavus* include oil seeds, ground nuts, maize, and dairy products, and, concomitantly, these foods induce the highest levels of aflatoxin production (79). Although aflatoxins appear to have received a disproportionate amount of the attention in mycotoxin research, many other toxic metabolites produced by conidial fungi are known, and new ones are continually being discovered (41,93,96,113).

In addition to *Aspergillus flavus*, several other moulds capable of inducing mycotoxicoses in animals and in humans have been recognized as allergenic fungi (Table 5–1). Outbreaks of mycotoxicoses are usually associated with fungal contamination of a specific food or feed. Whether or not low levels of toxic metabolites may also be released to the respiratory and/or digestive tracts during repeated exposure to airborne conidia has yet to be determined. Several allergenic fungi listed in Table 5–1 are also recognized as human pathogens. The most notable of these are *A. fumigatus, A. flavus, A nidulans, A. terreus,* and *Geotrichum candidum,* all potentially causative agents of pulmonary mycoses. In each case the infectious propagule that finds access to the respiratory tract is the airborne conidium. Typically these and most other pathogenic fungi are opportunistic, taking advantage of the debilitated and compromised host. Austwick (2) has distinguished respiratory allergies from infections by implying that the fungal spore germinates in only the latter. However, he recognized that this distinction is not biologically clear, and the two conditions may be closely related. Rippon (115) has pointed out that

> The ability of fungi to cause disease appears to be an accidental phenomenon . . .. The survival and growth of fungi at the elevated temperature of the body, the reduced oxidation-reduction environment of tissue, and the ability to overcome the host's defense mechanisms set apart these few species from the great numbers of saprophytic and plant pathogenic fungi.

At present we recognize approximately 175 pathogens among more than 100,000 species of fungi described (28). Although less than 30% of these pathogens have been seriously considered to be causative agents of cutaneous, severe subcutaneous, and systemic infections, a long and growing list of opportunistic fungi has been acknowledged and these fungi may colonize almost any tissue substrate of the debilitated patient and cause disease (115). Many fungi on this list are also recognized as potential allergenic fungi and have not been included in Table 5–1.

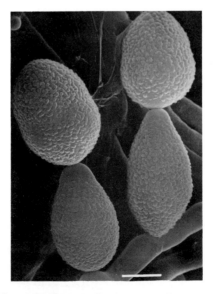

Fig. 5–2. Conidia of *Alternaria alternata*. Bar represents 5 μm.

Fig. 5–3.   Conidia and fertile branches of conidiophore of *Botrytis cinerea*. Bar represents 10 μm.

SIZE AND TOPOGRAPHY OF AIRBORNE CONIDIA

It has been estimated that 20% of the human population are atopic and easily sensitized by air spora at concentrations of $10^6/m^3$ or less (85). Such individuals demonstrate immediate-type allergic responses with hay fever-like symptoms when the upper respiratory tract is exposed to aeroallergens. As previously noted, when fungal conidia are involved in such types of immediate allergy, their size is generally greater than 5 μm and, therefore, include commonly occurring species such as *Alternaria alternata* (Fig. 5–2), *Botrytis cinerea* (Fig. 5–3), *Cladosporium herbarum* and *C. macrocarpum* (Fig. 5–4), *Drechslera* spp., and *Epicoccum purpurascens* (Fig. 5–5). Conidia of *Aspergillus* and *Penicillium* may also be included, although they are typically smaller than 5 μm in diameter and enter the airways as chains rather than as individual cells (Fig. 5–6).

Once access has been attained, either through the nasal or buccal cavity, conidial deposition depends on anatomic features (57) and on

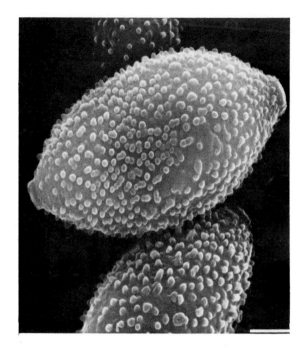

Fig. 5–4. Conidia of *Cladosporium macrocarpum*. Bar represents 2 μm.

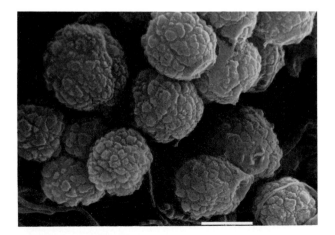

Fig. 5–5. Conidia of *Epicoccum purpurascens*. Bar represents 10 μm.

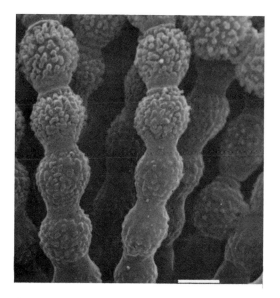

Fig. 5–6.   Chains of conidia produced by *Aspergillus penicilloides*. Bar represents 2 μm.

air-flow patterns within the respiratory tract, as well as on the size, density, and shape of the fungal propagules. Large conidia entering through the nares first encounter a network of nasal hairs that act as a mechanical barrier on which impaction of the spores may occur (2). Actually, however, many conidia manage to penetrate farther into the nasal cavity but then encounter the mucous lining of the nasal labyrinth, which efficiently removes most large particles from the air (58). The ciliated cells lining the nasal tissue sweep the mucus and entrapped conidia proximally toward the pharynx and the foreign particles are then removed to the digestive tract. Austwick (2) has concluded that deposition in the mucoid layer (Fig. 5–7) occurs

> because of the inertia of the spore carrying it on after a change in direction of air flow. This may take place in the upper respiratory tract around the nasal turbinates, at the far end of the nasal cavity where the air takes a downward turn to the trachea. . . . [as well as in lower regions] at bifurcations of the various grades of bronchi, and at places where turbulence speeds up the air near to the walls of the tract and hence facilitates inertial deposition.

Retention of large conidia in these lower airways is usually the result of bypassing the nasal filtration system during mouth inhalation (Fig. 5–1). Because the air velocity in the respiratory tract steadily decreases deeper within the system of airways (Fig. 5–7), sedimentation by gravity is also responsible for deposition of conidia. Surface topography of the conidium, in addition to size and density, plays a pivotal role in determining sites of deposition. The larger echinulate conidia of *Aspergillus*

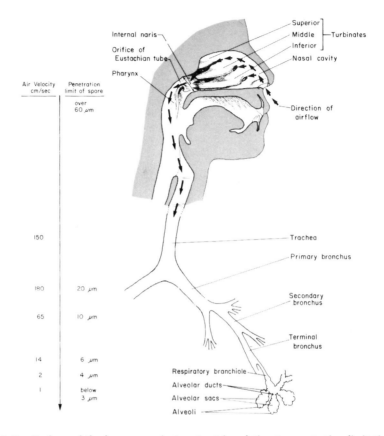

Air Velocity cm/sec | Penetration limit of spore

over 60 μm

150

180 | 20 μm

65 | 10 μm

14 | 6 μm

2 | 4 μm

1 | below 3 μm

Internal naris

Orifice of Eustachian tube

Pharynx

Superior
Middle ⎤ Turbinates
Inferior ⎦

Nasal cavity

Direction of airflow

Trachea

Primary bronchus

Secondary bronchus

Terminal bronchus

Respiratory branchiole

Alveolar ducts

Alveolar sacs

Alveoli

Fig. 5–7. Regions of the human respiratory tract in relation to penetration limitations of airborne spores and air velocity. (Modified from Austwick (2), Mitchell (95), and Tortora and Anagnostakos (137).)

Fig. 5–8. Conidia of *Aspergillus glaucus* (=*A. repens*). Bar represents 2 μm.

glaucus (= *A. repens;* Fig. 5–8) and *A. niger* (see Fig. 5–15) would be expected to impact higher in the respiratory tract than the smaller, smooth to slightly rough-surfaced conidia of *A. terreus* and *A. fumigatus* (Table 5–1). In fact, these latter conidia (<5 μm) are capable of reaching the respiratory bronchioles and alveolar ducts (<1 mm diameter) and even the alveoli (chambers approximately 0.5 mm in diameter; Fig. 5–7). These deep penetrating conidia are typically responsible for delayed allergy (85).

Conidia reaching these recesses of the respiratory tract are deposited mainly by sedimentation during the pause of a second at each inhalation (58). Because the walls of the alveolar sacs are not ciliated, conidia must be removed by phagocytic cells or may remain in the lungs for long periods. Those that germinate and ramify by hyphal penetration of tissue, or convert to a yeast phase and subsequently spread hematogenously and/or by lymphatic circulation, have become adapted to this new and hostile environment and pose a formidable threat to the well-being of the host, especially the atopic asthmatic. Schuyler et al. (122) have stated that patients with atopic dermatitis and asthma may be more susceptible to fungal infections and may exhibit depressed, delayed skin test reactivity to both recall and neoantigens. In a comparative assessment of the function of alveolar macrophages obtained by bronchoalveolar lavage from asthmatics and normal subjects, Godard et al. (51) concluded that phagocytic cells from the former showed impairment of both viability and functional activity.

CONIDIAL SURFACE ULTRASTRUCTURE

The initial and perhaps most significant interactions between host and conidium, either pathogenic or allergenic, occur at the cell surface, and knowledge of the nature of these surfaces is crucial to our understanding of the mechanisms of pathogenicity and allergenicity (108,126). Such interactions are particularly important in the case of immediate-type allergic response to the presence of fungal cells in the respiratory tract, because the conidium probably remains structurally intact and the host is, therefore, not exposed to cytoplasmic components of the inhaled propagule. Initial interactions between the conidium and host cells may involve host recognition of factors that are present on the conidial surface or released from the wall of the fungal cell. Investigations of the surface morphology and wall chemistry of allergenic fungi, combined with examinations of the immunologic activity of purified and characterized wall fractions, represent a logical approach to this problem. Although much information in this area is available for saprophytic and pathogenic yeasts, (e.g., 8,9,19,30,31,47,81,82,109,112,136,138), much less is known of the mycelial fungi (e.g., 12,14,27,40,52,61,62,64,65,120).

Results of studies of the surface ultrastructure of airborne conidia have

Fig. 5—9. Shadow replica of outermost surface of *Aspergillus niger* conidia. The replica was prepared by transferring dry conidia from the plate culture directly onto a clean surface (mica) and then shadowing in a vacuum evaporator. Bar represents 1 μm. (F, rodlet fascicles.)

commonly revealed a fibrous network of wall components, referred to as "rodlets" (Fig. 5–9) (32,50,66,67). These structures occur as interdigitated bundles, or fascicles, that give the appearance of a woven complex of wall fibrils. Among the conidial fungi, rodlets appear to be restricted to the surface of propagules and fertile hyphae (conidiogenous cells), and have not been demonstrated on the wall of vegetative hyphae (32). Morphologically similar wall constituents have been seen on spores produced by other groups of fungi (23,38,59,69,144) as well as on bacterial spores of *Bacillus, Streptomyces,* and *Micropolyspora* (1,22,45,54,70,127,134,143,145,146). Reports on the chemical composition of rodlets have been conflicting (see below).

In most cases the rodlet layer of conidia can be revealed by direct shadow replica preparations of dry cells (33), indicating that rodlets are exposed on the outermost surface of the cell envelope (34). In certain conidial fungi, however, the outermost surface of the propagule is instead comprised of a thin amorphous layer (Fig. 5–10), below which lies the rodlet complex (35,148). In such cases the rodlet layer can be exposed by freeze-fracturing and the interface between the amorphous and rodlet layers can be exposed by subsequent freeze-etching (49).

Differences also exist with respect to the ease of isolating rodlets from conidia (43). For example, certain strains of *Aspergillus niger* (35,37) and *Trichophyton mentagrophytes* (148) that require harsh mechanical and chemical treatments to release the rodlets from the cell envelope, while

Fig. 5–10.   Replica of freeze-fractured and deep-etched conidium of *Cladosporium macrocarpum*. The arrowheads indicate the interface between the outermost amorphous and rodlet layers; the arrow shows rodlets exposed through the thin amorphous layer (see Fig. 5–4). Bar represents 0.5 μm. (From Cole, G.T., et al. (35).)

conidia of other strains readily shed their rodlets when cell suspensions are simply agitated in water (Fig. 5–11). Although the microconidia of *T. mentagrophytes* in Figure 5–11 lack the outermost amorphous layer reported by Wu-Yuan and Hashimoto (148) for their strain, this deficiency is apparently not solely responsible for such variations in facility of rodlet isolation. Strains of *A. fumigatus* with exposed rodlet fascicles have been seen that demonstrate this same degree of variation. Conidia of the wild strain of *Neurospora crassa* examined by Dempsey and Beever (43) released sheets of rodlet fascicles when suspended in water and shaken vigorously by hand. The authors estimated that individual rodlets were 10 nm wide, and their visible length varied from about 35 to 240 nm.

Considerable variation exists in the length of rodlets and width of rodlet fascicles among conidial fungi (see Figs. 5–9. and 5–10). For example, in those species of *Penicillium* examined (66), rodlet length was seen to vary from 10 to 400 nm and fascicle width from 90 to 600 nm. On the other hand, patterns of rodlet fascicles on the surface of conidia appear to be similar for a particular genus (59). Individual rodlets revealed by shadow replicas at high magnification generally appear to consist of linear arrays of particles approximately 5 nm in diameter which

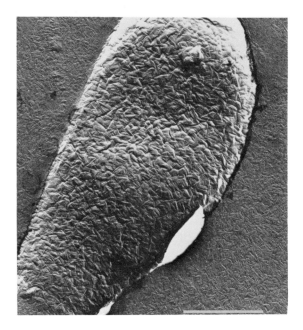

Fig. 5–11. Shadow replica of *Trichophyton mentagrophytes* microconidium. The replica was prepared by briefly suspending conidia from the plate culture in distilled water and then drying a droplet of the suspension on a clean surface (mica), followed by shadowing in a vacuum evaporator. Note the large number of rodlet fascicles that have been released from the conidial surface. Also note the lack of an outermost amorphous layer (see Fig. 5–10). Bar represents 1 µm.

form bands with a center-to-center spacing of approximately 8 to 10 nm between the rodlets (39,66).

At present, isolation and purification of rodlet subunits have not been achieved. Negatively stained rodlets of *N. crassa,* on the other hand, do not reveal a linear arrangement of particles but rather suggest that they are hollow fibers (43). Individual rodlets of the bacterial spores of *Micropolyspora angiospora* examined by negative staining appear to be a double-stranded helix, and it has been suggested that they constitute a helical polymer composed of globular subunits (134). The fine structure of rodlets, their arrangement on the surface of fungal and bacterial spores (i.e., structural association between individual rodlets as well as rodlet fascicles), and their relationship to enveloping and/or underlying wall components are fertile subjects for future research.

Beever and Dempsey (16) examined a mutant strain of *Neurospora crassa* that differed from the wild type in that its conidia were easily wetted, did not disperse readily in air currents, and lacked rodlets on their surface. The genetic difference was based on a single gene (mutant gene, *eas)* whose chromosomal locus has been established (104,124). The authors suggested that the rodlet layer contributes significantly to the hy-

Fig. 5–12.   Shadow replica of Nonidet P-40-treated conidium of *Aspergillus niger,* showing apparently swollen rodlet fascicles (see Fig. 5–9). (From Cole, G.T., and Pope, L.M. (37).)

drophobicity of conidia and to their dispersal efficiency in air. If hydrophobic conidia of *Aspergillus niger* with exposed rodlet fascicles are incubated in 1% Nonidet P-40, a nonionic detergent (37), the cells are wetted while the rodlet layer remains essentially intact (Fig. 5–12). Only slight swelling of the fascicles and separation of adjacent rodlets are detected, probably due to hydration of the cell wall (see Fig. 5–9). The viability of detergent-treated conidia is the same as that of the control. Analysis of the protein content of the cell-free supernatant (Fig. 5–13), using a modified Lowry procedure (44), revealed that a significant amount of protein is released during incubation (37). The amount of released protein was reduced as the concentration of detergent was decreased, and varied according to the nature of the wetting agent employed. Because the rodlet layer was not removed during hydration and the conidia remained viable, the protein released may represent solubilized wall components that have diffused through the rodlet layer. It has been suggested that the detergent is responsible for breakdown of the hydrophobic barrier, which may not entirely be attributed to the rodlets alone, as well as for increasing the porosity of the cell wall (42) and subsequently releasing solubilized components of presumed wall origin through the rodlet layer.

Of particular interest is that wall hydration and concomitant release of protein into the supernatant also occurred when conidia were incu-

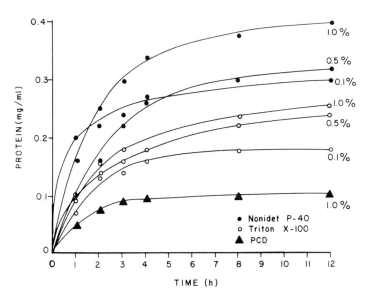

Fig. 5–13. Protein released from conidia of *Aspergillus niger* during incubation in varying concentrations (0.1%, 0.5%, 1.0%) of nonionic detergents (Nonidet P-40 and Triton X-100) and 1.0% L-α-phosphatidylcholine dipalmitoyl (PCD) over a 12-hour period at 25° C with constant agitation (200 rpm on a New Brunswick shaker-incubator). Protein determinations were performed using a modified Lowry method (44). (From Cole, G.T., and Pope, L.M. (37) and Cole, G.T. (34).)

bated in 1.0% L-α-phosphatidylcholine dipalmitoyl, a commercial lung surfactant (Fig. 5–13), although proteins were detected in much lower concentrations than when nonionic detergents were employed. This incubation mixture was not buffered, however, and the procedure was therefore not performed under optimal conditions. The implication is nevertheless intriguing: how important is the interaction between hydrophobic components of the fungal cell wall and lung surfactant during initial recognition and host response to inhaled conidia? The biochemical nature of these interactions, as well as those between other mucous secretions (20) and allergenic conidia, has not been investigated as yet.

WALL CHEMISTRY

Three different procedures have commonly been used in the preparation of fungal walls for combined chemical and ultrastructural analyses: cell fractionation, wall purification, and chemical-ultrastructural examinations of wall components (35); selective enzyme digestion (72); and protoplast production followed by examination of progressive wall regeneration and reversion to normal morphology (101,102). Our conidial wall isolation procedures have primarily involved controlled mechanical disruption using ultrasonication and fractionation techniques (35,40),

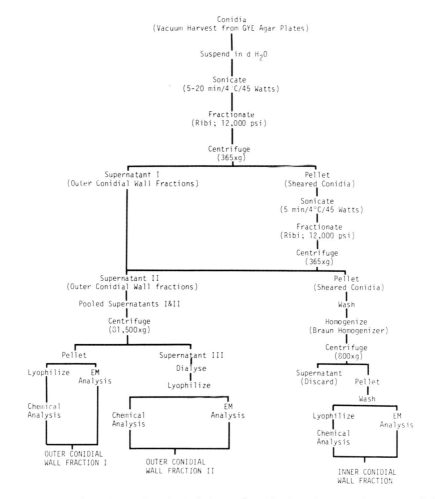

Fig. 5–14. Outline of procedure for isolation and purification of outer and inner conidial wall fractions. (From Cole, G.T., et al. (40).)

although advantages clearly exist in combining all three procedures for structural and chemical analyses of cell walls.

An outline of our preparatory procedure is presented in Figure 5–14. Conidia are vacuum-harvested from agar plates and resuspended in 200 ml of distilled water (4° C) or in a suitable detergent for subsequent analysis of solubilized products in the cell-free supernatant. Aliquots of 50 ml of the conidial suspension are then sonicated (one to four times, 5 min at 4° C) using a Branson Sonifier (Branson Sonic Power Company, Danbury, Connecticut), set at 45 watts. The pooled sonicated suspension is then passed through a Ribi cell fractionator (Sorvall Model RF-1, refrigerated) at a pressure of 12,000 psi. This apparatus, which is a modified

French press, permits conidia to be sheared but not ruptured at well-controlled pressure and temperature (5–10° C).

The high degree of efficiency of the shearing process is demonstrated by comparing Figure 5–15A and B. The sheared intact cells (Fig. 5–15B) are separated from the wall suspension by centrifugation. The supernatant contains the outer conidial wall fraction released during sonication and fractionation (Fig. 5–15C). The pellet (sheared conidia) is resuspended in 200 ml of distilled water (4° C), sonicated, fractionated, and centrifuged as above. The supernatants containing the outer conidial wall fractions are pooled while the pellet (sheared conidia) is washed and then homogenized in a Braun homogenizer (Model MSK, Bronwill Scientific Inc., Rochester, New York) using 0.45- to 0.50-mm glass beads during continuous cooling with liquid CO_2. The homogenized suspension is immediately centrifuged and the pellet (inner conidial wall, Fig. 5–15D) is washed with distilled water (4° C) at least five times.

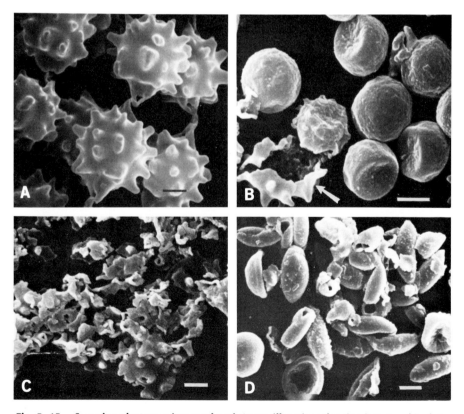

Fig. 5–15. Scanning electron micrographs of *Aspergillus niger* showing intact, echinulate conidia *(A)*, sheared conidia after passage through the Ribi cell fractionator *(B; arrow indicates sheared outer wall layer)*, isolated and purified outer conidial wall layer *(C; see Table 5–3)*, and isolated and purified inner conidial wall layer *(D; see Table 5–3)*. Bars represent 1 μm. (From Cole, G.T., et al. (35).)

In our recent investigations of the chemical and antigenic composition of the walls of the pathogenic fungus *Coccidioides immitis* (36), the outer conidial wall fraction was separated by centrifugation into a high-speed pellet (85,500 xg, 2 hours) and a supernatant. The latter (outer conidial wall fraction II) was subsequently dialyzed for 24 hours against four changes of distilled water (4° C). All wall fractions were examined by electron microscopy and lyophilized in preparation for chemical analysis.

Results of these analyses of conidial wall fractions of *A. niger* are presented in Table 5–3. As discussed earlier, nonionic detergent treatment released substantial amounts of protein and lipids from the cell wall but did not inhibit conidial germination. When these same detergent-treated conidia were sonicated, wall components were removed from the cell surface (Fig. 5–16). Rodlets, which comprise the outermost wall layer of this strain of *A. niger*, were dislodged as large fascicles. Below and apparently attached to the rodlet fascicles is a pebbly layer, visible in Figure 5–16. This layer is morphologically similar to the hyphal wall component of *A. niger* and *A. awamori* described by Bobbitt et al. (21) and identified as nigeran: a hot, water-soluble, linear, alternating $(1 \rightarrow 3)$, $(1 \rightarrow 4)$-α-D-glucan (10,11). Below this pebbly layer is the distinct fibrous wall material also shown in Figure 5–16 which was demonstrated to be alkali-soluble and was interpreted as β-$(1 \rightarrow 3)$-glucan (37).

The chemical composition of the conidial wall fractions derived by sonication of detergent-treated cells is presented in Table 5–3. If these data are compared to those for the composition of the outer wall layer, obtained from a distilled water suspension of conidia sheared in the Ribi cell fractionator (Figs. 5–15B and C, 5–17), clear differences are recognized, especially with respect to protein content. The composition of the inner wall fraction of *A. niger* obtained after homogenization of the sheared conidia (Fig. 5–15B and D), differs from that of both the outer and detergent wall fractions (Table 5–3). The inner fraction has a much higher neutral carbohydrate content but lower protein content than the other wall fractions. Using this procedure of selective wall shearing and cell fractionation, we have begun to understand the distribution of components in the conidial wall. The surface of the conidium consists primarily of hydrophobic proteins and lipid, although the composition of the rodlets themselves is still controversial. Rodlets may be composed of a polysaccharide complex, as in *Streptomyces coelicolor* (127), a crystallized S-glucan as in *Schizophyllum commune* (144), a primarily proteinaceous component as in conidia of *Trichophyton mentagrophytes* and *Neurospora crassa* (17,60,148), a lipoprotein, as in *A. niger* (37), or a glycoprotein with a molecular weight of 12,000, as in *Syncephalastrum racemosum* (69). It is quite possible that differences exist in rodlet composition between major groups of microorganisms. However, our experience with *A. niger* is that it is difficult to purify rodlets. Even in those strains that shed their rodlets readily, agitation of conidial suspensions results in some disruption of the cell surface and release of sheets of

TABLE 5–3. **Summary of Chemical Composition of Conidial Wall Fractions of *Aspergillus niger***

	NP-40 Detergent-Soluble Fraction (%)[a]	Sonicated Outer Wall Layer (%)[a]	Sheared Outer Wall Layer (%)[b]	Homogenized Inner Wall Layer (%)[b]
Total Carbohydrate	3.5	6.9	2.5	31.2
Mannose	(15.6)[c]	(30.8)	(31.5)	(29.5)
Galactose	(48.7)	(44.2)	(24.2)	(30.7)
Glucose	(35.7)	(25.0)	(37.5)	(36.4)
Total hexosamine	nd[d]	3.3	2.2	6.5
N-acetylglucosamine	nd	nd	(0.1)[h]	(0.2)[h]
Peptides	33.6[e]	34.9[e]	63.4[i]	20.0[i]
	(38.0)[f]	(35.6)[f]		
		(38.4)[g]		
Lipids				
Readily extractable	17.0	20.0	16.2	11.0
Bound	15.5	10.0	8.6	23.5
Ash	1.5	nd	1.6	nm[j]
Phosphorus	nd	nd	0.1	0.1
Recovery	71.1	75.1	94.6	92.3

[a]Data from Cole and Pope (37). Sonicated outer wall layer obtained from NP-40 detergent-treated conidia.
[b]Data from Cole et al (35). Sheared outer layer (prepared by Ribi cell fractionation) and homogenized inner layer (prepared by Bronwill glass bead homogenization) obtained from aqueous conidial suspension.
[c]Monosaccharide content expressed as a percentage of total carbohydrate.
[d]Not determined.
[e]Determined by ninhydrin method.
[f]Based on adjusted total nitrogen.
[g]Determined by summation of amino acids.
[h]N-acetylglucosamine/total hexosamine.
[i]Determined by the Lowry method.
[j]Not measurable.

intact fascicles (43) that are not totally free of underlying and interstitial wall components. These contaminating wall components could significantly influence results of chemical analysis. Similarly, alkali digestion of the cell surface during rodlet isolation (37,69) does not break down the rodlet fascicles, and may remove additional underlying wall components. In *A. niger*, melanin is a major fraction of the conidial wall (133) but is not necessarily part of the rodlet complex. Freeze-fractures of hyaline conidia

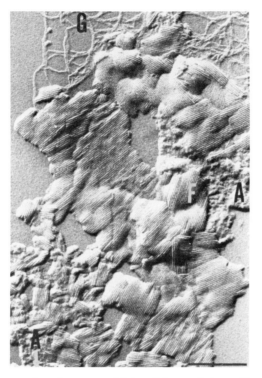

Fig. 5–16.   Shadow replica of outer conidial wall layer of *Aspergillus niger* obtained by sonication of detergent-treated (Nonidet P-40) conidia (see Table 5–3). (A, amorphous, pebbly wall layer; G, fibrous glucan; F, rodlet fascicle.) Bar represents 0.5 μm. (From Cole, G.T., and Pope, L.M. (37).)

produced by *Geotrichum candidum* and *Cladobotryum varium* also demonstrate rodlets (39,97).

Our interpretation of *A. niger* is that the rodlet fascicles shown in thin section in Fig. 5–18 are firmly bound to an underlying layer composed mainly of the melanin complex, protein, and lipid. These three components in dematiacious species contaminate rodlet preparations irrespective of the isolation procedure employed (Fig. 5–19). Below this layer is a region of the conidial wall composed primarily of fibrous and amorphous glucan in addition to protein, which probably occurs mainly as glycoprotein. It is important to emphasize here that different wall layers actually occur as gradations rather than as distinct zonations. When the outer wall layer of *A. niger* conidia is removed in the Ribi cell fractionator (Fig. 5–15B), the zone of shearing appears to be the interface between the outer layers and this glucan-glycoprotein layer. The composition of the inner region of the glucan layer that borders on the innermost chitin-containing region is unknown. Results of enzyme dissections of the hyphal walls of *Neurospora crassa* suggest that a discrete layer of protein separates the glucans and glycoproteins from the chitinous region, with chitin microfibrils embedded

Fig. 5–17.   Shadow replica of outer conidial wall layer of *Aspergillus niger* obtained by Ribi cell fractionation (see Fig. 5–15 *B,C)* of an aqueous suspension of conidia. The rodlet fascicles are intact and show an original interdigitated pattern (see Table 5–3). Bar represents 0.5 µm.

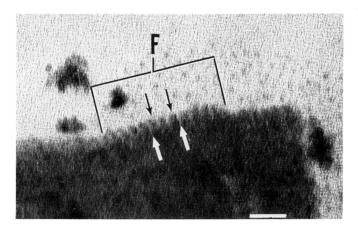

Fig. 5–18.   Thin section of outer conidial wall of *Aspergillus niger* obtained by Ribi cell fractionation (see Fig. 5–17) and then subjected to chitinase digestion. The black arrows indicate tangentially sectioned rodlets; the white arrows indicate electron-translucent material between rodlets. (F, rodlet fascicle.) Bar represents 20 nm. (From Cole, et al. (35).)

Fig. 5–19.   Shadow replica of rodlets *(arrows)* and rodlet fascicles (F) of *Aspergillus niger* isolated from conidia treated with 0.1 N NaOH (12 hours, 25° C). Bar represents 1 μm.

principally in proteinaceous material (26, 53). Similar studies of *A. niger* conidia using techniques of selective enzyme dissection combined with ultrastructural analyses should provide further clarification of the nature of the conidial wall.

FUTURE RESEARCH

The airborne conidia have not received adequate attention beyond the areas of epidemiologic and aerobiologic research. We have discussed some natural substrates and habitats of conidial fungi that provide sources of air-dispersed propagules. Morphologic features of the conidium have been considered in relation to possible sites of deposition (impaction or sedimentation) within the human respiratory tract. Based on results of electron microscopic examinations and related chemical analyses of the conidial wall, the cell envelope was discussed from the standpoint of its being a reservoir of potential allergens, a theme that forms the basis of our concluding discussion of future research.

In their consideration of the role of airborne fungal spores in the production of respiratory allergy, Salvaggio and Aukrust (117) pointed out that "the true clinical impact of airborne fungi on respiratory diseases, particularly on bronchial asthma, is still largely unknown, and the characterization, purification and standardization of allergens from a wide range of fungal species will be necessary before this type of information can be obtained." Areas for concerted research efforts, therefore, include the development of improved and standardized procedures for preparing and analyzing mould extracts and refining reliable testing procedures for determining the potency of allergenic extracts.

Current approaches to the identification and purification of clinically relevant antigens from cell walls of the pathogenic moulds *Coccidioides immitis* (36) and *Aspergillus fumigatus* (29,61,63,64,65) may also be applicable to allergenic conidial fungi. Emphasis was initially placed on developing simple procedures for obtaining antigens from cell walls of these fungi grown in culture, which would provide standardized reproducible reagents for serologic diagnosis. Ward et al. (141) have demonstrated that an alkali- and water-soluble antigen extracted from the mycelial phase cell walls of *C. immitis* elicited positive reactions in 92% of *Coccidioides*-sensitized guinea pigs, whereas only 54% of the same guinea pigs reacted to commercial coccidioidin (125). The latter, a toluene-induced whole cell lysate of *C. immitis*, is still used extensively as a serodiagnostic reagent in clinical laboratories. It contains multiple antigens (74) that are involved in differing immunologic responses. Hearn and Mackenzie (64) have prepared wall-located antigens from *Aspergillus fumigatus* mycelia by extraction with the nonionic detergent Triton X-100. They compared the antigenic composition of this extract to that of a water-soluble preparation of homogenized mycelia by the use of two-dimensional immunoelectrophoresis (2D-IEP). The latter procedure as well as crossed radioimmunoelectrophoresis (CRIE) is a useful technique for the analysis of complex antigenic mixtures (4,5,87,142). Use of *in vitro* serologic systems has demonstrated that wall antigens of *Aspergillus fumigatus* have three to eight times the sensitivity of the water-soluble preparations. In contrast to the complex pattern of precipitin peaks observed when the water-soluble fraction reacts with hyperimmune sera, significantly fewer peaks were seen when wall-located antigens were used in the 2D-IEP system. Although the wall preparation still consisted of a mixture of antigens, Hearn and Mackenzie concluded that recurrence of a limited number of antigen-antibody precipitin arcs that "override batch and strain variability could be important in a study of pathogenicity and in the induction of an immune response"(64). Such simplification of preparatory procedures and apparent reproducibility of antigenic content are positive aspects of this approach. However, Huppert (73) has cautioned that it is important to determine the degree of cross reactivity between comparable antigen preparations from closely related fungi. Such cross reactivity is known to occur among fungal pathogens (75) but has not been well explored in the allergenic fungi (117). Another related problem of interpretation is that immunologic reactions may be either a response to a single antigen or a composite of responses to several antigens, some of which are shared by several species (46). In regard to studies of fungal pathogenicity, Huppert (73) has also pointed out that "evaluation of the immunologic reactivity of an antigen preparation must not be divorced from the type of clinical disease but must be integrated into the entire picture of patient infection and response."

The current emphasis in medical mycology is on understanding the intricate mechanisms by which host cells learn to recognize, remember, and respond to a foreign insult containing the many antigens of an infectious organism. This will require isolation of highly purified, preferably monomolecular, antigens. The preparation of such antigens is now possible, and is based on the following hypothesis: monoclonal antibody formation against a specific antigen within a mixture (e.g., wall extract) can be used to purify that antigen by solid phase immunoadsorption, and the specific antigen can then be recovered by elution with dissociating agents (121).

Such procedures, although still in their early stages of application to fungi, may revolutionize the preparation of serologic diagnostic reagents for the mycoses. Baldo et al. (7) have recognized the value of such approaches to the standardization of allergens. Once an allergen has been isolated in pure form and its biologic activity has been unequivocally established, the way is clear for the production of a monoclonal antibody standard. These same authors indicated that immediate introduction of monoclonal antibody standards is possible for such purified and biologically characterized allergens as ragweed, rye grass, and timothy grass pollens. Characterization of fungal allergens is not yet at this point. However, the application of simple conidial wall extraction procedures for obtaining allergenic preparations, combined with the technology for analyzing their content, examining their immunologic reactivity and selecting and purifying clinically relevant allergens, are promising areas for future research.

REFERENCES

1. Aronson, A.I., and Fitz-James, P.: Structure and morphogenesis of the bacterial spore coat. Bacteriol. Rev., **40**:360, 1976.
2. Austwick, P.K.C.: The role of spores in the allergies and mycoses of man and animals. *In* The Fungus Spore. Edited by M.F. Madelin. London, Butterworths, 1966.
3. Ayerst, G.: The effects of moisture and temperature on growth and spore germination in some fungi. J. Stored Prod. Res., **5**:127, 1969.
4. Axelsen, N.H. (ed.): Quantitative Immunoelectrophoresis. Oslo, Universitetsforlaget, 1975.
5. Axelsen, N.H., Kroll, J., and Weeke, B. (eds.): A Manual of Quantitative Immunoelectrophoresis. Oslo, Universitetsforlaget, 1973.
6. Baggerman, W.I.: Heat resistance of yeast cells and fungal spores. *In* Introduction to Food-Borne Fungi. Edited by R.A. Samson, E.S. Hoekstra, and C.A.N. van Oorschot. Baarn, Netherlands, Centraalbureau voor Schimmelcultures, 1981.
7. Baldo, B.A., et al.: A unified approach to the standardization of allergens. Med. J. Aust., **2**:651, 1980.

8. Ballou, C.: Structure and biosynthesis of the mannan component of the yeast cell envelope. Adv. Microbiol. Physiol., **14**:93, 1976.

9. Ballou, C.E., Lipke, P.N., and Raschke, W.C.: Structure and immunochemistry of the cell wall mannans from *Saccaromyces chevalieri, Saccharomyces italicus, Saccharomyces diastaticus,* and *Saccharomyces carlsbergensis.* J. Bacteriol., **117**:461, 1974.

10. Barker, S.A., et al.: Studies of *A. niger.* VI. The separation and structures of oligosaccharides for nigeran. J. Chem. Soc., 2448, 1956.

11. Barker, S.A. Bourne, E.J., and Stacey, M.: Studies of *Aspergillus niger.* I. The structure of the polyglucosan synthesized by *Aspergillus niger* 152. J. Chem. Soc., 3084, 1953.

12. Barreto-Berter, E., Gorin, P.A.J., and Travassos, L.R.: Cell constituents of mycelia and conidia of *Aspergillus fumigatus.* Carbohydr. Res., **95**:205, 1981.

13. Barron, G.L.: The Genera of Hyphomycetes from Soil. Baltimore, Williams & Wilkins, 1968.

14. Bartnicki-Garcia, S., and Lippman, E.: Fungal cell wall composition. *In* Handbook of Microbiology. 2nd Ed. Vol. IV. Edited by A.I. Laskin and H. Lechevalier. Cleveland, CRC Press, 1981.

15. Baseman, J.B., et al.: Mucus and surfactant synthesis and secretion by cultured hamster respiratory cells. Environ. Hlth. Perspect., **35**:139, 1980.

16. Beever, R.E., and Dempsey, G.P.: Function of rodlets on the surface of fungal spores. Nature, **272**:608, 1978.

17. Beever, R.E., Redgwell, R.J., and Dempsey, G.P.: Purification and chemical characterization of the rodlet layer of *Neurospora crassa* conidia. J. Bacteriol., **140**:1063, 1979.

18. Berrens, L.: The Chemistry of Atopic Allergens. Basel, S. Karger, 1971.

19. Bhattacharjee, A.K., Kwon-Chung, K.J., and Glandemans, C.P.J.: On the structure of the capsular polysaccharide from *Cryptococcus neoformans* serotype C-II. Mol. Immunol., **16**:531, 1979.

20. Boat, T.F., and Cheng, P.W.: Biochemistry of airway mucus secretions. Fed. Proc., **39**:3067, 1980.

21. Bobbitt, T.F., et al.: Distribution and conformation of crystalline nigeran in hyphal walls of *Aspergillus niger* and *Aspergillus awamori.* J. Bacteriol., **132**:691, 1977.

22. Bradley, S.G., and Ritzi. D.: Composition and ultrastructure of *Streptomyces venezuelae.* J. Bacteriol., **95**:2358, 1968.

23. Bronchart, R., and Demoulin, V.: Ultrastructure de la paroides basidiosporés de *Lycoperdon* et de *Scleroderma* (Gastéromycètes) comparée à celle de quelques autres spores de champignons. Protoplasma, **72**:179, 1971.

24. Bronswijk, J.E.M.H. v.: House Dust Biology for Allergists, Acarologists and Mycologists. Netherlands, NIB, 1981.

25. Burge, H.A., Solomon, W.R., and Boise, J.R.: Microbial prevalence in domestic humidifiers. Appl. Environ. Microbiol., **39**:840, 1980.

26. Burnett, J.H.: Aspects of the structure and growth of hyphal walls. *In* Fungal Walls and Hyphal Growth. Edited by J.H. Burnett and A.P.J. Trinci. London, Cambridge University Press, 1979.

27. Catley, B.J.: The extracellular polysaccharide, pullulan produced by *Aureobasidium pullulans:* A relationship between elaboration rate and morphology. J. Gen. Microbiol., **120**:265, 1980.

28. Chandler, F.W., Kaplan, W., and Ajello, L.: Histopathology of Mycotic Diseases. Chicago, Year Book Medical Publishers, 1980.

29. Chaparas, S.D., et al.; Characterization of antigens from *Aspergillus fumigatus.* V. Reactivity in immunodiffusion tests with serums from patients with aspergillosis caused by *Aspergillus flavus, A. niger, and A. fumigatus.* Am. Rev. Resp. Dis., **122**:647, 1980.

30. Chattaway, F.W., Holmes, M.R., and Barlow, A.J.E.: Cell wall composition of the mycelial and blastospore forms of *Candida albicans.* J. Gen. Microbiol., **51**:367, 1968.

31. Cherniak, R., et al.: Structure and antigenic activity of the capsular polysaccharide of *Cryptococcus neoformans* serotype A. Mol. Immunol., **17**:1025, 1980.

32. Cole, G.T.: A correlation between rodlet orientation and conidiogenesis in Hyphomycetes. Can. J. Bot., **51**:2413, 1973.

33. Cole, G.T.: Techniques for examining developmental and ultrastructural aspects of

conidial fungi. *In* Biology of Conidial Fungi. Vol. 2. Edited by G.T. Cole and B. Kendrick. New York, Academic Press, 1981.
34. Cole, G.T.: Infectious fungal propagules. *In* Microbiology—1983. Edited by D. Schlessinger. Washington, D.C., Amer. Soc. Microbiol., 1983.
35. Cole, G.T., et al.: Surface ultrastructure and chemical composition of the cell walls of conidial fungi. Exp. Mycol., **3**:132, 1979.
36. Cole, G.T., and Huppert, M.: Wall chemistry and ultrastructure in cell types of *Coccidioides* (Abstract). Annu. Meeting Am. Soc. Microbiol., 330, 1982.
37. Cole, G.T., and Pope, L.M.: Surface ultrastructure of *Aspergillus niger* conidia. *In* The Fungal Spore: Morphogenetic Controls. Edited by G. Turian and H.R. Hohl. London, Academic Press, 1981.
38. Cole, G.T., and Ramirez-Mitchell, R.: Comparative scanning electron microscopy of *Pencillium* conidia subjected to critical point drying, freeze-drying and freeze-etching. Scanning Electron Microscopy 1974, **II**:367, 1974.
39. Cole, G.T., and Samson, R.A.: Patterns of Development in Conidial Fungi. London, Pitman, 1979.
40. Cole, G.T., Sun, S.H., and Huppert, M.: Isolation and ultrastructural examination of conidial wall components of *Coccidioides* and *Aspergillus*. Scanning Electron Microscopy **1982**:1677, 1982.
41. Cole, R.J., and Cox, R.H.: Handbook of Toxic Fungal Metabolites. N.Y., Academic Press, 1981.
42. Cope, J.E.: The porosity of the cell wall of *Candida albicans*. J. Gen. Microbiol., **119**:253, 1980.
43. Dempsey, G.P., and Beever, R.E.: Electron microscopy of the rodlet layer of *Neurospora crassa* conidia. J. Bacteriol., **140**:1050, 1979.
44. Dulley, J.R., and Grieve, P.A.: A simple technique for eliminating interference by detergents in the Lowry method of protein determination. Anal. Biochem., **64**:136, 1975.
45. Ebersold, H.R., et al.: A freeze-substitution and freeze-fracture study of bacterial spore structures. J. Ultrastruct. Res., **76**:71, 1981.
46. Ehrhard, H.B., and Pine, L.: Factors influencing the production of H and M antigens by *Histoplasma capsulatum*. Effect of physical factors and composition of medium. Appl. Microbiol., **23**:250, 1972.
47. Elinov, N.P., et al.: Mannan produced by *Rhodotorula rubra* strain 14. Carbohydr. Res., **75**:185, 1979.
48. Evans, H.C.: Thermophilous fungi of coal spoil tips. II. Occurrence, distribution and temperature relationships. Trans. Br. Mycol. Soc., **57**:255, 1971.
49. Fineran, B.: Freeze-etching. *In* Electron Microscopy and Cytochemistry of Plant Cells. Edited by J.L. Hall. Amsterdam, Elsevier, 1978.
50. Ghiorse, W.C., and Edwards, M.R.: Ultrastructure of *Aspergillus fumigatus* conidia development and maturation. Protoplasma, **76**:49, 1973.
51. Godard, P., et al.: Functional assessment of alveolar macrophages: Comparison of cells from asthmatics and normal subjects. J. Allergy Clin. Immunol., **70**:88, 1981.
52. Gomez-Miranda, B., and Leal, J.A.: Extracellular and cell wall polysaccharides of *Aspergillus alliaceus*. Trans. Br. Mycol. Soc., **76**:249, 1981.
53. Gooday, G.W., and Trinci, A.P.J.: Wall structure and biosynthesis in fungi. *In* The Eukaryotic Microbial Cell. Edited by G.W. Gooday, D. Lloyd, and A.P.J. Trinci. London, Cambridge University Press, 1980.
54. Gould, G.W., Stubbs, J.M., and King, W.L.: Structure and composition of resistant layers in bacterial spore coats. J. Gen. Microbiol., **60**:347, 1970.
55. Grappel, S.F., Bishop, C.T., and Blank, F.: Immunology of dermatophytes and dermatophytosis. Bacteriol. Rev., *38*:222, 1974.
56. Gravesen, S.: Fungi as a cause of allergic disease. Allergy, **34**:135, 1979.
57. Greenwood, M.F., and Holland, P.: The mammalian respiratory tract surface. A scanning electron microscopic study. Lab. Invest., **27**:296, 1972.
58. Gregory, P.H.: The Microbiology of the Atmosphere. 2nd Ed. Aylesbury, England, Leonard Hill, 1973.
59. Hallett, I.C., and Beever, R.E.: Rodlets on the surface of *Neurospora* conidia. Trans. Br. Mycol. Soc., **77**:26, 1981.
60. Hashimoto, T., Wu-Yuan, C.C., and Blumenthal, H.J.: Isolation and characterization

of the rodlet layer of *Trichophyton mentagrophytes* microconidial wall. J. Bacteriol., **127**:1543, 1976.

61. Hearn, V.M., et al.: Preparation of *Aspergillus fumigatus* antigens and their analysis by two-dimensional immunoelectrophoresis. J. Med. Microbiol., **13**:451, 1980.
62. Hearn, V.M., and Mackenzie, D.W.R.: The preparation and chemical composition of fractions from *Aspergillus fumigatus* wall and protoplasts possessing antigenic activity. J. Gen. Microbiol., **112**:35, 1979.
63. Hearn, V.M., and Mackenzie, D.W.R.: Antigenic activity of subcellular fractions of *Aspergillus fumigatus*. Zentralbl. Bakteriol. Parasitkde., (suppl.) **8**:173, 1980.
64. Hearn, V.M., and Mackenzie, D.W.R.: Analysis of wall antigens of *Aspergillus fumigatus* by two-dimensional immunoelectrophoresis. J. Med. Microbiol., **14**:119, 1981.
65. Hearn, V.M., Proctor, A.G., and Mackenzie, D.W.R.: The preparation and partial characterization of antigenic fractions obtained from the mycelial walls of several *Aspergillus* species. J. Gen. Microbiol., **119**:41, 1980.
66. Hess, W.M., Sassen, M.M.A., and Remsen, C.C.: Surface characteristics of *Penicillium* conidia. Mycologia, **60**:290, 1968.
67. Hess, W.M., and Stocks, D.L.: Surface characteristics of *Aspergillus* conidia. Mycologia, **61**:560, 1969.
68. Hobot, J.A., and Gull, K.: Changes in the organization of surface rodlets during germination of *Syncephalastrum racemosum* sporangiospores. Protoplasma, **107**:339, 1981.
69. Hobot, J.A., and Gull, K.: Structure and biochemistry of the spore surface of *Syncephalastrum racemosum*. Curr. Top. Microbiol. Immunol., **5**:183, 1981.
70. Holt, S.C., and Leadbetter, E.R.: Comparative ultrastructure of selected aerobic spore-forming bacteria: A freeze-etch study. Bacteriol. Rev., **33**:346, 1969.
71. Hořejší, N., and Šach, J.: A syndrome resembling farmer's lung in workers inhaling spores of *Aspergillus* and *Penicillium* moulds. Thorax, **15**:212, 1960.
72. Hunsley, D., and Burnett, J.H.: The ultrastructural architecture of the walls of some hyphal fungi. J. Gen. Microbiol., **62**:203, 1970.
73. Huppert, M.: Antigens used for measuring immunolgical reactivity. *In* Pathogenicity and Detection. Vol. IIA. Edited by D.H. Howard, New York, Marcel Dekker, 1983.
74. Huppert, M., et al.: Antigenic analysis of coccidioidin and spherulin determined by two-dimensional immunoelectrophoresis. Infect. Immunol., **20**: 541, 1978.
75. Huppert, M., et al.: Common antigens among systemic disease fungi analysed by two-dimensional immunoelectrophoresis. Infect. Immunol., **23**:479, 1979.
76. Huppert, M., et al.: The propagule as an infectous agent in coccidioidomycosis. *In* Microbiology—1983. Edited by D. Schlessinger. Washington, DC, Am. Soc. Microb., 1983.
77. Ishii, A., et al.: Mite fauna and fungal flora in house dust from homes of asthmatic children. Allergy, **34**:379, 1979.
78. Jayaprakash, K.B., Rati, E., and Ramalingam, A.: *Aspergillus flavus* in the air of working environments. Curr. Sci., **47**:920, 1978.
79. Jones, B.D.: Aflatoxins and related compounds: Occurrence in foods and feeds. *In* Mycotoxic Fungi, Mycotoxins and Mycotoxicoses. Vol. I. Edited by T. Wyllie and L. Morehouse. New York, Marcel Dekker, 1977.
80. Jong, S.C.: Isolation, cultivation, and maintenance of conidial fungi. *In* Biology of Conidial Fungi. Vol. 2. Edited by G.T. Cole and B. Kendrick. New York, Academic Press, 1981.
81. Käppeli, O., Müller, M., and Fiechter, A.: Chemical and structural alterations at the cell surface of *Candida tropicalis*, induced by hydrocarbon substrate. J. Bacteriol., **133**:952, 1978.
82. Karson, E.M., and Ballou, C.E.: Biosynthesis of yeast mannan. Properties of a mannosylphosphate transferase in *Saccharomyces cerevisiae*. J. Biol. Chem., **253**:6485, 1978.
83. King, R.J.: The surfactant system of the lung. Fed. Proc., **33**:2238, 1974.
84. King, R.J.: Utilization of alveolar epithelial type II cells for the study of pulmonary surfactant. Fed. Proc., **38**:2637, 1979.
85. Lacey, J.: The aerobiology of conidial fungi. *In* Biology of Conidial Fungi. Vol. 1. Edited by G.T. Cole and B. Kendrick. New York, Academic Press, 1981.
86. Last, J.A., and Kaizu, T.: Mucus glycoprotein secretion by tracheal explants: Effects of pollutants. Environ. Hlth. Perspect., **35**:131, 1980.

87. Laurell, C.B.: Antigen-antibody crossed electrophoresis. Anal. Biochem., **10**:358, 1965.
88. Lidwell, O.M.: Mikroorganismer: Levende stuf i luften. *In* Termisk og Atmosfaerisk Indeklima. Edited by N. Jonassen. Lyngby, Denmark, Polyteknisk Forlag, 1970.
89. Lustgraaf, B.: Xerophilic fungi in mattress dust. Mykosen, **20**:101, 1977.
90. Lustgraaf, B.: Ecological relationships between xerophilic fungi and house-dust mites (Acarida: Pyroglyphidae). Oecologia, **33**:351, 1978.
91. Lustgraaf, B., and Bronswijk, J.E.H.M.V.: Fungi living in house dust. Ann. Allergy, **39**:152, 1977.
92. Lustgraaf, B., and Jorde, W.: Pyroglyphid mites, xerophilic fungi and allergenic activity in dust from hospital mattresses. Acta Allergol., **32**:406, 1977.
93. Mislivec, P.B.: Mycotoxin production by conidial fungi. *In* Biology of Conidial Fungi. Vol. 2. Edited by G.T. Cole and B. Kendrick. New York, Academic Press, 1981.
94. Mislivec, P.B., and Tuite, J.: Temperature and relative humidity requirements of species of *Penicillium* isolated from yellow dent corn kernels. Mycologia, **62**:75, 1970.
95. Mitchell, R.I.: Retention of aerosol particles in the respiratory tract. A review. Am. Rev. Resp. Dis., **82**:627, 1960.
96. Northolt, M.D., and Soentoro, P.S.S.: Fungal growth on foodstuffs related to mycotoxin contamination. *In* Introduction to Food-Borne Fungi. Edited by R.A. Samson, E.S. Hoekstra, and C.A.N. van Oorschot. Baarn, Netherlands, Centraalbureau voor Schimmelcultures, 1981.
97. Nozawa, Y., Kasai, R., and Cole, G.T.: Ultrastructure and chemistry of cell walls of *Geotrichum candidum:* With special reference to conidiogenesis. Jpn. J. Med. Mycol., **22**:202, 1981.
98. Panasenko, V.T.: Ecology of microfungi. Bot. Rev., **33**:189, 1967.
99. Papavizas, G.C.: Factors influencing invasion of stored wheat seed by *Aspergillus* spp. and the effects of such invasion on germination and amount of germ damage. Diss. Abstr., **17**:2397, 1957.
100. Patterson, R.: Allergic Diseases: Diagnosis and Management. Philadelphia, J.B. Lippincott., 1972.
101. Peberdy, J.F.: Fungal protoplasts: Isolation, reversion and fusion. Ann. Rev. Microbiol., **33**:21, 1979.
102. Peberdy, J.F.: Wall biogenesis by protoplasts. *In* Fungal Walls and Hyphal Growth. Edited by J.H. Burnett and A.P.J. Trinci. London, Cambridge University Press, 1979.
103. Pennycook, S.R.: The air spora of Auckland City, New Zealand. I. Seasonal and dial periodicities. N.Z.J. Sci., **23**:27, 1980.
104. Perkins, D.D., et al.: Chromosomal loci of *Neurospora crassa*. Microbiol. Rev., **46**:426, 1982.
105. Pitt, J.I.: Xerophilic fungi and the spoilage of foods of plant origin. *In* Water Relations of Foods. Edited by R.B. Duckworth. New York, Academic Press, 1975.
106. Pitt, J.I.: The Genus *Penicillium* and its Teleomorphic States *Eupenicillium* and *Talaromyces*. London, Academic Press, 1979.
107. Pitt, J.I.; Food spoilage and biodeterioration. *In* Biology of Conidial Fungi. Vol. 2. Edited by G.T. Cole and B. Kendrick. New York, Academic Press, 1981.
108. Raa, J.: Cell surface biochemistry related to specificity of pathogenesis and virulence of microorganisms. *In* Biology and Chemistry of Eukaryotic Cell Surfaces. Edited by E.Y.C. Lee and E.E. Smith. New York, Academic Press, 1974.
109. Raizada, M.K., Schutzbach, J.S., and Ankel, H.: *Cryptococcus laurentii* cell envelope glycoprotein. Evidence for separate oligosaccharide side chains of different composition and structure. J. Biol. Chem., **250**:3310, 1975.
110. Raper, K.B., and Fennell, D.I.: The Genus *Aspergillus*. Baltimore, Williams & Wilkins, 1965.
111. Reid, L.M., and Jones, R.: Mucous membrane of respiratory epithelium. Environ. Hlth. Perspect., **135**:113, 1980.
112. Reiss, E., et al.: Structural analysis of mannans from *Candida albicans* serotypes A and B and from *Torulopsis glabrata* by methylation gas chromatography mass spectrometry and exo-α-mannanase. Biomed. Mass Spectrom., **8**:252, 1981.
113. Reisz, J.: Mykotoxine in Lebensmitteln. Stuttgart, Gustav Fischer Verlag, 1981.
114. Rijckaert, G.: Fast-Releasing Allergens from some Organisms Living in House Dust. Ph. D. thesis, Univ. Nijmegen, Netherlands, 1981.

115. Rippon, J.W.: Medical Mycology: The Pathogenic Fungi and the Pathogenic Acti-nomycetes. 2nd Ed. Philadelphia, W.B. Saunders, 1982.
116. Salvaggio, J., et al.: New Orleans asthma. II. Relationship of climatologic and seasonal factors to outbreak. J. Allergy, **45**:257, 1970.
117. Salvaggio, J., and Aukrust, L.: Mold-induced asthma. J. Allergy Clin. Immunol., **68**:327, 1981.
118. Salvaggio, J., and Klein, R.: New Orleans asthma. I. Characterization of individuals involved in epidemics. J. Allergy, **39**:227, 1967.
119. Samson, R.A., Hoekstra, E.S., and van Oorschot, C.A.N. (eds.): Introduction to Food-Borne Fungi. Baarn, Netherlands, Centraalbureau voor Schimmelcultures, 1981.
120. San-Blas, G.: The cell wall of fungal human pathogens: Its possible role in host-parasite relationships. Mycopathol., **79**:159, 1982.
121. Scharff, M.D., Roberts, S., and Thammana, P.: Monoclonal antibodies. J. Infect. Dis., **143**:346, 1981.
122. Schuyler, M.R., et al.: Corticosteroid-sensitive lymphocytes are normal in atopic asthma. J. Allergy Clin. Immunol., **68**:72, 1981.
123. Scott, W.J.: Water relations of food spoilage microorganisms. Adv. Food Res., **7**:83, 1957.
124. Selitrennikoff, C.P.: Easily-wettable, a new mutant. *Neurospora* Newsletter, **23**:23, 1976.
125. Smith, C.E., et al.: The use of coccidioidin. Am. Rev. Tuberc., **57**:330, 1948.
126. Smith, H.: Microbial surfaces in relation to pathogenicity. Bacteriol. Rev., **41**:475, 1977.
127. Smucker, R.A., and Pfister, R.M.: Characteristics of *Streptomyces coelicolor* A$_3$(2) aerial spore rodlet mosaic. Can. J. Bot., **24**:397, 1978.
128. Solomon, W.R.: Assessing fungus prevalence in domestic interior J. Allergy Clin. Immunol., **56**:235, 1975.
129. Staib, F.: Mycoses caused by fungal spores in indoor air. Zentrabl. Bakteriol. Hyg. Abt. I, **176**:142, 1982.
130. Staib, F., et al.: A comparative study of antigens of *Aspergillus fumigatus* isolates from patients and soil of ornamental plants in the immuno diffusion test. Zentrabl. Bak-teriol. Hyg. Abt. I, **242**:93, 1978.
131. Staib, F., et al.: *Aspergillus fumigatus* and *Aspergillus niger* in two potted ornamental plants, cactus *(Epiphyllum truncatum)* and clivia *(Clivia miniata).* Biological and epi-demiological aspects. Mycopathol., **66**:27, 1978.
132. Staib, F., et al.: *Aspergillus fumigatus* in der Topferde von zimmerplanzen. Bundes-gesundheitsblatt **21**:209, 1978.
133. Swan, G.A.: Structure, chemistry and biochemistry of the melanins. *In* Progress in the Chemistry of Organic Natural Products. Vol. 31. Edited by W. Herz, H. Grisebach, and G.W. Kirby, New York, Springer, 1974.
134. Takeo, K.: Existence of a surface configuration on the aerial spore and aerial mycelium of *Micropolyspora.* J. Gen. Microb., **95**:17, 1976.
135. Tansey, M.R., and Brock, T.D.: Microbial life at high temperatures: Ecological aspects. *In* Microbiological Life in Extreme Environments. Edited by D.J. Kushner. New York, Academic Press, 1978.
136. Thieme, T.R., and Ballou, C.E.: Nature of the phosphodiester linkage of the phos-phomannan from the yeast *Kloeckera brevis.* Biochem., **19**:4121, 1971.
137. Tortora, G.J., and Anagnostakos, N.P.: Principles of Anatomy and Physiology. 2nd Ed. San Francisco, Canfield, 1978.
138. Tronchin, G., et al.: Cytochemical and ultrastructural studies of *Candida albicans.* II. Evidence for a cell wall coat using Concanavalin A. J. Ultrastruct. Res., **75**:50, 1981.
139. Tronchin, G., Poulain, D., and Biguet, J.: Études cytochemiques et ultrastructurales de la paroi de *Candida albicans.* I. Localisation des mannanes par utilisation de Con-canavaline A sur coupes ultrafines. Arch. Microbiol., **123**:245, 1979.
140. Velcovsky, H.G., and Graubner, M.: Allergische Alveolites durch Inhalation von Schimmelpilzsporen aus Blumenerde. Med. Wochenschr., **106**:115, 1981.
141. Ward, E.R., et al.: Delayed-type hypersensitivity responses to a cell wall fraction of the mycelial phase of *Coccidioides immitis.* Infect. Immunol., **12**:1093, 1975.
142. Weeke, B., and Lowenstein, H.: Allergens identified in crossed radioimmunoelec-trophoresis. Scand. J. Immunol., **2**:149, 1973.

143. Wehrli, E., Scherrer, P., and Kübler, O.: The crystalline layers in spores of *Bacillus cereus* and *Bacillus thuringiensis* studied by high resolution electron microscopy. Eur. J. Cell Biol., **20**:283, 1980.
144. Wessels, J.G.H., et al.: Chemical and morphological characterization of the hyphal wall surface of the basidiomycete *Schizophyllum commune*. Biochim. Biophys. Acta, **273**:346, 1972.
145. Williams, S.T., et al.: Fine structure of the spore sheath of some *Streptomyces* species. J. Gen. Microbiol., **72**:249, 1972.
146. Williams, S.T., Sharples, G.P., and Bradshaw, R.M.: Spore formation in *Actinomadura dassonvillei* (Brocq-Ronssean) Lechevalier and Lechevalier. J. Gen. Microbiol., **84**:415, 1974.
147. Wolf, H.W., et al.: Sampling Microbiological Aerosols. Washington, D.C., U.S. Government Printing Office, U.S. Department of Health, Education, and Welfare, P.H.S. Monog. No. 60, 1959.
148. Wu-Yuan, C., and Hashimoto, T.: Architecture and chemistry of microconidial walls of *Trichophyton mentagrophytes*. J. Bacteriol., **129**:1584, 1977.

Chapter 6

MOULD ALLERGENS

Donald R. Hoffman

Fungal spores are ubiquitous constituents of both indoor and outdoor air samples. Some genera of airborne fungal spores such as *Alternaria* and *Cladosporium* are found throughout most of the world. Spores are generally considered to be important causes of both allergic rhinitis and allergic asthma; mould allergy, however, is the least understood and studied of the major forms of inhalant allergy (35). Pollen allergy has been extensively researched; a great deal is now known about the allergen content of a few pollens and the human immune response to these allergens. The most extensively studied aeroallergen is ragweed pollen *(Ambrosia elatior)*, from which at least six distinct allergens have been isolated and purified (27). Only in the past few years have investigators attempted to isolate and characterize allergens from imperfect fungi. Results of some of these recent studies have shown that in many cases the particular problems of working with fungi make this task even more difficult than with other allergen sources.

Fungi are quite different from other aeroallergens. The inhaled particle consists of entire living cells. Some fungi (mainly pathogenic) have two phases: mould, consisting of spores and mycelia, and yeast. Airborne fungal particles are mostly spores rather than mycelial fragments. Most common genera of fungi have a large number of species, and there are distinct strains within most of the species. In addition there are multiple antigenically related genera. Problems arise even when working with a well-characterized strain because of the inherent ability of the organism to vary in response to growth conditions. For some fungi it is almost impossible to grow two consecutive cultures with even similar antigenic contents (2). Many fungal allergens are glycopeptide in nature and may share common antigenic and allergenic determinants with closely related and relatively unrelated organisms. In addition, many fungi are known to produce substances such as C-reactive substance that can combine nonspecifically with immunoglobulin molecules.

Of the many genera of imperfect fungi, only a few have been investigated in any detail with respect to allergen content. These include the two most important allergenic moulds, *Alternaria* and *Cladosporium*, as well as *Aspergillus*, *Monilia*, *Dreschlera* (often erroneously called *Helminthosporium*), *Candida*, and members of the *Basidiomycetes*. Progress in the isolation and characterization of fungal allergens has been slow and has been further complicated by the finding that many allergens exhibit substantial microheterogeneity with respect to both molecular weight and isoelectric point. Recent advances in techniques for separation of macromolecules that can be used in microgram amounts, however, as well as the production of fungal spores in quantity with relatively little mycelia present, should lead to significant advances in our knowledge of fungal allergens in the next decade.

STUDYING MOULD ALLERGENS

Cultures

It is now possible to grow many allergenically important fungi on totally synthetic defined media, which are free of macromolecules (15,29). After a strain has been grown for several passes in a synthetic medium it should be maximally stable. Defined media eliminate the problems of separating fungal constituents from medium constituents or from other irrelevant antigens. For many fungi sporulation can be prevented by growing the culture in liquid medium with constant agitation; spores can be best produced on solid media. The Murashige-Skoog plant salt and minimal organic mixture (29) has been tested in our laboratories and been found to support good growth, with abundant sporulation of many species of imperfect fungi. In many cases sporulation seemed to be strain-dependent using both defined and traditional media. In liquid cultures, it is necessary to maintain strict control over pH, oxygen level, carbon dioxide level, and concentration of metabolic products, as well as temperature and agitation rate.

Extraction

Control over extraction conditions is essential in the production of extracts with reproducible allergen content. The pH, buffer, and salt content of the extraction fluid are important, as are the temperature and duration of extraction. Excessively long extraction periods, particularly at room temperature or above, may lead to destruction of allergens by hydrolytic enzymes present in the extract. Ball milling or other vigorous grinding methods appear to be of little value, because the increased protein content is found mainly in the structural components of the cell.

In some cases this structural material causes precipitation of much of the extract. With most moulds, dialysis or ultrafiltration, with a cutoff of 10,000 daltons, does not cause any significant loss of allergenic activity and in many cases enhances activity (e.g., for RAST testing; see below). Because the physiologic extraction of pollen allergens in the nose occurs between 25° and 37° C at a pH of about 7.4 in a solution of normal saline containing albumin, this is probably a good starting point. Albumin should not be included in the extraction fluid for the preparation of purified allergens. In our experience, 50% glycerin appears to stabilize extracts against hydrolysis and is recommended for the preparation of extracts to be used for skin testing.

Testing for Allergenic Activity

In most instances the simplest method for determining whether or not an extract or fraction contains allergenic activity is by direct skin testing using allergic subjects. Because most extracts contain multiple allergens and reactivity to individual allergens varies among patients, it is best to use a panel of volunteers and also to include two control groups to evaluate nonspecific reactivity: nonatopics, and subjects allergic to pollen allergens only. Either prick testing or intradermal titration can be used. Both methods require great care to be reproducible. Intradermal titration is about two logs (10^2) more sensitive and is usually more quantitative than prick testing.

In vitro methods for assessing allergenic activity are usually more convenient. Various antigens can be tested *in vitro* and the patient's reactivity can be compared at different times by using frozen serum specimens. The most basic *in vitro* test for assay of allergen activity is the radioallergosorbent test (RAST) (48). Allergen is chemically bound to an insoluble polymer support, usually cellulose or agarose, that has been treated with cyanogen bromide at a pH of 10 to 11. The allergen polymer is incubated with patient serum, washed, incubated with radioactively labeled antihuman IgE, and washed again. The radioactivity bound to the polymer is then determined, and the activity is calculated from a calibration curve (13). The results are not strictly quantitative because of interference from IgG antibodies. In the case of crude and partially purified allergens, insufficient quantities of some allergens bind to the polymer. Several modifcations of the RAST have been used to obtain quantitative data (52).

Activities of allergens can be compared by inhibition of RAST (14). A reference allergen is used to prepare the polymer. Patient sera are mixed with varying dilutions of the allergens to be tested, usually before adding the mixture to the polymer. RAST is then run and the results are determined as percentages of binding without adding inhibitor. The calculated results are then plotted semilogarithmically. The ratio of the 50% inhibition points of the test sample to the reference sample is a measure of potency, and the similarity of slopes indicates similarity of the allergens. RAST inhibition should be a powerful technique, but the results are often difficult to interpret.

The principle of the RAST can be combined with various separation techniques to assess allergenic activity of various components of a mixture. The most powerful of these techniques is crossed radioimmunoelectrophoresis (CRIE) (24). The mixture to be studied is first electrophoresed in agarose. Gels containing precipitating antibodies raised against the components being studied or against the entire mixture are poured on both sides of the first separation gel. The plate is then turned 90° and subjected to electrophoresis again, producing parabola-shaped precipitates. After extensive washing and air drying, the plate is incubated with human reaginic serum overnight, washed again, and incubated with radiolabeled human IgE. After another thorough washing and drying, an autoradiograph is obtained from the plate and compared to the original plate, which has been stained with a protein stain. This technique has high resolving power and has been used extensively in investigations of *Alternaria* and *Cladosporium* allergens.

Direct challenge of allergic patients by inhalation of small doses of extract has been used in many studies of inhalant allergens (30). Although useful for pollen studies its use with moulds is controversial, because mould allergens cause not only allergic asthma but also late phase reactions, some of which are similar to those in bronchopulmonary aspergillosis (42). Because monitoring of the subject is required for at least 4 to 6 hours after completion of the tests, this technique has been little used by most investigators. Even though they are relatively safe to perform with mould allergens, nasal and conjunctival challenge tests are less quantifiable than are bronchial inhalation challenges.

Techniques for the Separation and Study of Mould Allergen Extracts

Several texts have been published with details of methods that can be used for studying mould allergens (9,22,47,49). Macromolecules of the allergens can be fractionated by size using gel filtration. Various media are currently available that complement the use of the traditional cross-linked dextran gels. Aqueous gel filtration media for high-performance liquid chromatography (HPLC) have been developed recently that allow rapid studies with very small amounts of extract. The use of reversed phase HPLC has not been reported with mould extracts, but is worthy of investigation.

Ion exchange chromatography, the traditional method for separating molecules by charge, is most useful in second-stage purification, usually following gel filtration. The newer technique of chromatofocusing separates molecules that are relatively charge-homogeneous. In this method the proteins are bound to an ion exchanger at a pH above their isoelectric points and then eluted with a mixture of amphoteric buffers that generate a gradient of decreasing pH. Results of preliminary studies with *Alternaria* extracts show this to be a very useful technique. In our experience most of the macro-molecules in mould extracts are relatively acidic, although there are small amounts of basic and neutral proteins that appear to have allergenic activity.

Electrophoretic techniques are also very useful for characterizing fungal extracts. Supports for preparative electrophoresis include starch gel, agarose, pevikon, and polyacrylamide gel. Molecular weights and subunit compositions of proteins are commonly determined from electrophoresis of reduced and sodium dodecyl sulfate (SDS)-denatured samples. Molecules may be separated by their isoelectric points using isoelectric focusing or electrophoresis in a natural pH gradient. The analytic form of this technique is usually carried out in thin layers of polyacrylamide or special agarose gels; the preparative method is carried out in a bed of specially prepared cross-linked dextran gel.

Isoelectric focusing is combined with SDS-polyacrylamide gel electrophoresis to give a two-dimensional mapping technique that can resolve thousands of components (32). These maps are often made from biosynthetically radiolabeled extracts, because of the relative insensitivity of most protein stains. The introduction of silver staining methods with over a tenfold increase in sensitivity, however, (28,36) have made radiolabeling unnecessary in many cases.

Electrophoresis can be combined with immunologic techniques to yield various methods for separating and characterizing allergen mixtures (46). The best known method is immunoelectrophoresis. If antigen is electrophoresed into an antibody-containing gel, the technique is known as rocket electrophoresis. The combination of electrophoresis in the first dimension and rocket electrophoresis in the second is called crossed immunoelectrophoresis (CIEP). Incorporation of additional antigen or antibody into intermediate gels multiplies the information derived from CIEP.

Affinity chromatography is a useful method for isolating molecules that exhibit charge and size heterogeneity. Many enzymes can be purified by absorption to and elution from insolubilized molecules that resemble substrates or co-factors. A substance can be purified by adsorption to an antibody specific for that substance that has been chemically attached to an insoluble support (e.g., agarose particles). The antigens can then be eluted from the antibody by using acid buffers, chaotropic ions, or hapten solutions. Similar methods can be used for purifying proteins with hydrophobic binding sites or dye binding sites. Lectins and complexing agents such as phenylboronate are used for purifying carbohydrates. Affinity techniques have been very useful in studies of mould allergens (15).

DIFFERENCES AMONG SPORES, MYCELIA, AND YEAST FORMS OF FUNGI

The first demonstration of phase-specific antigens in fungi was reported by Syverson et al. (44). Using CIEP, they identified six yeast-specific and six mycelium (pseudohyphae)-specific antigens in *Candida albicans*. This study was followed up by Manning and Mitchell (26), who combined cross-absorption methods with O'Farrell's two-dimensional electrophoresis (32). They proposed that many of the differences between yeast and mycelial-phase maps were due to protein modification, and that few if any new proteins were synthesized during phase conversion.

In the only report comparing *Alternaria* mycelial antigens to spore antigens, Hoffman et al. (15) found approximately eight antigens in spore extracts that could not be found in mycelial extracts. This spore-specific fraction was prepared by immune absorption and shown to contain at least one highly allergenic component by both RAST and skin testing. More recent data indicate that one of the spore-specific antigens is a protein having an isoelectric point of 6.2 to 6.3. This protein could not be found in mycelial extracts by any of various immunologic and physicochemical techniques. These findings may be important because spores, not mycelia, are the natural aeroallergen. It does appear, however, that the great majority of allergens are common to both spores and mycelia. From results of preliminary studies of other genera, including *Cladosporium*, it appears that most antigens are common to both spores and mycelia.

Previously, some investigators claimed that mould extracts made from pellicle (mycelia) were inferior in diagnostic efficiency to those made from either dialyzed medium or a combination of pellicle and medium (5,33,34,37). Extracts prepared in our laboratory from culture supernate, however, contain relatively little allergenic activity as compared to that of the mycelial or spore extracts. Results of studies using CIEP and isoelectric focusing have shown that media used for growing *Alternaria* contain only a portion of the mycelial antigens found in extracts. Some proteins such as beta-galactosidase (23) and some other hydrolases are secreted into the medium during growth.

Alternaria alternata

Spores of *Alternaria alternata* and those of the closely related genera *Stemphylium* and *Ulocladium* are the most important mould allergens in the United States (41,43). *Alternaria* is a major allergen on the East coast, in the upper Midwest, and in southern California, probably ranking third only to ragweed and grass pollens as a natural cause of allergy. There are 14 different strains of *Alternaria alternata* listed in the American

Type Culture Catalogue (8), and many others are held in laboratories throughout the world. Because of its importance, *Alternaria* is the most studied of mould allergens.

Schumacher and Jeffery (40) have studied the variation between culture filtrates from different isolates and also from different batches of the same isolate. They found substantial biochemical differences, but only little difference in reaction with rabbit antibodies. Using radiolabeled antigen, it was demonstrated that most of the *Alternaria* antigens contained carbohydrates and there was extensive cross reaction with *Stemphylium* and *Curvularia* (39). Extensive cross reactivity was also reported between culture filtrates and mycelial extracts (38). Vijay et al. (45) have demonstrated that *A. tenuis (alternata)* exhibited extensive cross reactivity with *Alternaria solani,* and found their allergenic fractions to have similar properties.

Yunginger et al. (50,51) have shown that a major allergenic fraction of *Alternaria alternata* is a protein and carbohydrate containing molecular weight peak of greater than 10,000 daltons. They extracted *Alternaria* powder, fractionated the extract sequentially by ammonium sulfate precipitation, DEAE-cellulose ion exchange chromatography, preparative electrofocusing, and Sephadex G-100 gel filtration. The carbohydrate-rich material isolated was called ALT-1; it was found to be heterogeneous by polyacrylamide gel electrophoresis and analytic isoelectric focusing. The isoelectric points of the approximately 12 bands ranged from 4.0 to 4.5, and the broad peak eluted from the G-100 column at molecular weights corresponding to 25,000 to 50,000 daltons. Amino acid analysis of subfractions also revealed some heterogeneity. All subfractions were potent inhibitors of RAST to both crude *Alternaria* and to each other; again, heterogeneity was apparent. Use of ALT-1 in RAST showed a high specificity for *Alternaria*-sensitive asthmatics.

Allergens with very similar properties to ALT-1 have been reported by Lowenstein (24) and by Salvaggio and Aukrust (35), who found it in eight strains of *Alternaria alternata.* The Aukrust allergen has been studied by CRIE and the isoelectric points were reported to be from 4.30 to 4.65. Recently ALT-1 has been compared to the Scandinavian isolates and the fractions were found to be highly cross-reactive but with definite spur formation, indicating that they were not totally identical (31). Hoffman et al. (16) reported that ALT-1 activity is found in both spore and mycelial extracts.

A recent report by Budd et al. described the isolation from *Alternaria* pellicle of a basic peptide allergen having an isoelectric point of 9.5 to 9.8 and a molecular weight below 14,000 (6). This allergen contains at least two bands on SDS polyacrylamide gel electrophoresis. Studies using RAST in our laboratory have confirmed the existence of allergen activity in the most basic fraction of *Alternaria* spore and mycelial extracts.

Cladosporium spp.

Cladosporium is the most commonly found airborne fungal spore in most of the temperate zones of the world (43). Several species are important allergens, including *C. herbarum* and *C. cladosporoides*. Studying the allergenic composition of *C. herbarum* by the use of CRIE and other techniques, Aukrust has isolated two significant allergens (1–4,35). *C. herbarum* was found to contain over 60 distinct precipitable antigens by CIEP; less than half of these antigens, however, were shared by extracts of *C. herbarum* from two different sources. When a pool of sera from 35 allergic individuals was used, CRIE indicated four important allergens and more than ten other allergens. The allergens were not consistently present in all extracts examined.

The major allergen, named Ag-32 from the CRIE pattern, was isolated by gel filtration and by isoelectric focusing. Its molecular weight was approximately 13,000, with an isoelectric point of 3.4 to 4.4. Of 35 patients' sera, 23 reacted with Ag-32. This allergen was found to have five isoelectric variants, all of which are immunologically identical. Studies of extracts of ten isolates of *C. herbarum* showed Ag-32 contents of from 0 to 1000% of the reference extract. A second less important allergen, Ag-54, had a molecular weight of approximately 25,000 and an isoelectric point of 5.0. Ag-54 was reactive with 20% of the patients' sera.

In a collaborative study with Aukrust, we examined the Ag-32 and Ag-54 contents of spores and mycelia from three species of *Cladosporium*. *C. cladosporoides* spores contained a large amount of Ag-32, about 12 times the content of mycelial extract from the same species. *C. tenuisum* spores also contained more Ag-32 than mycelia. Ag-32 was present in our California isolate of *C. herbarum* mycelia, but it was not possible to obtain spores in quantity from this isolate. Only traces of Ag-54 were found in any of the extracts, and none was detectable in *C. cladosporoides* spores.

Aspergillus spp.

Extracts of various species of *Aspergillus* have been studied to produce diagnostic reagents for allergic bronchopulmonary aspergillosis (42) and to control toxin production (35). Dessaint et al. (11) used sera from 60 patients with both IgE and IgG antibodies to *Aspergillus fumigatus* to identify a specific allergen. They had previously reported a species-specific antigen, C2. The IgG antibodies from 32 of the patients reacted with C2, and from 1 to 14 antigen bands were found by immunoelectrophoresis using the 32 patients' sera. Very little IgE antibody reacted with C2. Using radioimmunoelectrophoresis they found a single strong allergenic component distinct from C2.

Kim and Chaparas (21) developed a reproducible method for growing *Aspergillus fumigatus* in totally synthetic medium and for preparing a consistent extract. Fractionation of the mycelial extract yielded a carbohydrate fraction devoid of IgE-binding activity, and also a complex series of protein fractions that possessed IgE-binding activity by RAST (20). One fraction, APIFB, was significantly more active in skin testing than crude extract; it contained at least 19 components by fused rocket electrophoresis. In further studies Chaparas et al. demonstrated the existence of species specificity in the genus *Aspergillus;* also, they found that mycelial extract is significantly more potent than culture medium filtrate (10).

Kauffmann and De Vries (18) also studied the growth conditions for preparing a standardized *A. fumigatus* extract. They found that 4 to 5 weeks of growth was optimum, and that the glucose concentration should be relatively low. In further studies they examined the binding of IgG antibodies from patients' sera using crossed immunoelectrophoresis (CIEP) and crossed immunoelectrofocusing. Many precipitin bands were observed, almost all with isoelectric points from 3 to 6. Malo et al. (25) examined six different extracts of *A. fumigatus,* including a "home-produced" strain by CIEP and by precipitation analysis using human sera. They found that no single serum reacted to all the extracts by either method. Calvanico et al. (7) reported the isolation of an antigen from cell sap of *A. fumigatus* that reacts (by precipitin test) with 75% of sera from patients with aspergilloma and allergic bronchopulmonary aspergillosis. This antigen consists of four polypeptide chains of 45,000 daltons linked through disulfide bonds, with an isoelectric point of 5.2 to 5.6. The molecules contained 12.5% neutral hexose. Related antigens of lower molecular weight, which may be partially degraded forms, were also found.

Other Genera

Yeast and mycelial antigens of *Candida albicans* have been investigated to prepare diagnostic reagents for detection of systemic candidosis (12,26,44). Greenfield and Jones (12) purified and characterized a cytoplasmic antigen from *C. albicans* that was released during invasive infection and that stimulated a strong antibody response in both humans and experimental animals. This antigen, which gave a single band in polyacrylamide gel electrophoresis, was a single polypeptide chain having a molecular weight of 54,300 daltons, and consisting of about 435 amino acid residues. The antigen contained about 5% carbohydrates and was acidic.

Little is known about the antigenic and allergenic composition of other allergenic fungi. Use of commercial extracts to study these is further complicated by misidentification of many of the original cultures. When slides of cultures used to produce commercial extracts from several manufacturers were examined, several common errors were found. The members of the genus *Helminthosporium* are generally obligate plant pathogens with limited hosts; *Helminthosporium* used to prepare commercial extracts was found to be *Dreschlera,* which is a common outdoor airborne fungus. One manufacturer prepared several types of extracts labeled as the same species; in one case, however, the *Stemphylium* was true *Stemphylium* and in another it was a distantly related genus. It is important to verify the identity of the isolate used for study and for extract preparation, and to identify strains recently isolated from the environments of allergic patients.

CROSS REACTIVITY AND SHARED DETERMINANTS

Apparent cross reactivity between related and unrelated moulds is a commonly observed phenomenon (17). Because actual cross inhibition studies have rarely been performed, the inference of cross reactivity is usually made from multiple reactivity in the same patient. Hoffman and Kozak (16) studied a group of 76 sera from mould-allergic patients by RAST inhibition. They demonstrated the expected cross reaction between *Alternaria, Stemphylium,* and *Curvularia,* and sometimes with *Spondylocladium* and *Dreschlera.* Allergens other than ALT-1 were important in these cross reactions. Patterns of positive reactions to these five dematiacious fungi varied from reactivity to only a single extract to reactivity with all five extracts. *Epicoccum* and *Fusarium* were shown to exhibit incomplete cross reactivity. Salvaggio and Aukrust (35) reported the existence of a single cross-reactive antigen between *Cladosporium herbarum* and *Alternaria alternata.*

Occasional patients are found who react with most mould extracts. Results of studies in our laboratory suggest that many of these patients have IgE antibodies that recognize common carbohydrate or glycopeptide determinants from various sources. In recent years many examples of similar phenomena have been reported, including cross inhibition among pollen, food, and venom allergens in RAST studies of sera from patients who recognize these common specificities. Definitive studies of cross reactivity require the use of standardized extracts or allergens and of carefully selected patient sera.

In summary, mould allergens are the least studied group of aeroallergens. The natural exposure is to the spore, but in many cases the mycelium appears to contain most of the same antigens and allergens as the spore. Commercial extracts are prepared from pellicle, which usually only contains a small amount of spore. Major allergens have been isolated from *Alternaria alternata* by several investigators. Some minor allergens of *Alternaria* have also been described. Several allergens have been isolated from *Cladosporium,* one of which is a major allergen. Knowledge of the allergenic content of other fungi is minimal. Limited studies suggest that there may be significant cross reactivity among various genera of moulds.

It is essential for investigators to work with carefully identified strains of fungi that are comparable to those isolated from the environment of the allergic patient. By carefully defining the media controlling growth and extraction conditions, pure potent extracts may be developed. These mould extracts can then be studied by the various techniques employed to purify and identify important allergens and to standardize their concentration.

REFERENCES

1. Aas, K., et al.: Immediate type hypersensitivity to common molds. Allergy, **35**:443, 1980.
2. Aukrust, L.: Cross radioimmunoelectrophoretic studies of distinct allergens in two extracts of *Cladosporium herbarum*. Int. Arch. Allergy Appl. Immunol., **58**:371, 1979.
3. Aukrust, L.: Allergens in *Cladosporium herbarum*. *In* Advances in Allergology and Immunology. Edited by A. Oehling. Oxford, Pergamon Press, 1980.
4. Aukrust, L., and Borch, S.M.: Partial purification and characterization of two *Cladosporium herbarum* allergens. Int. Arch. Allergy Appl. Immunol., **60**:68, 1979.
5. Browning, W.H.: Mold fungi in the etiology of respiratory allergic diseases. II. Mold extracts—a statistical study. J. Allergy, **14**:231, 1942.
6. Budd, T.W., et al.: The isolation of a unique basic peptide allergen from *Alternaria* extracts. J. Allergy Clin. Immunol. **69**:145, 1981.
7. Calvanico, N.J., et al.: Antigens of *Aspergillus fumigatus*. I. Purification of a cytoplasmic antigen reactive with sera of patients with *Aspergillus*-related disease. Clin. Exp. Immunol., **45**:662, 1981.
8. Catalogue of Strains, I.: The American Type Culture Collection. 12th Ed. Washington, D.C., 1976.
9. Catsimpoolas, N., and Drysdale, J.: Biological and Biomedical Applications of Isoelectric Focusing. New York, Plenum Press, 1977.
10. Chaparas, S.D., et al.: Characterization of antigens from *Aspergillus fumigatus*. V. Reactivity in immunodiffusion tests with serums from patients with aspergillosis caused by *Aspergillus flavus*, *A. niger*, and *A. fumigatus*. Am. Rev. Resp. Dis., **122**:647, 1980.
11. Dessaint, J.P., et al.: Serum concentration of specific IgE antibody against *Aspergillus fumigatus* and identification of the fungal allergen. Clin. Immunol. Immunopathol., **5**:314, 1976.
12. Greenfield, R.A., and Jones, J.M.: Purification and characterization of a major cytoplasmic antigen of *Candida albicans*. Infect. Immunol., **34**:469, 1981.
13. Hoffman, D.R.: The use and interpretation of RAST to stinging insect venoms. Ann. Allergy, **42**:224, 1979.
14. Hoffman, D.R.: Allergens in hymenoptera venom. VI. Cross reactivity of human IgE antibodies to the three vespid venoms and between vespid and paper wasp venoms. Ann. Allergy, **46**:304, 1981.
15. Hoffman, D.R., et al.: Isolation of spore-specific allergens from *Alternaria*. Ann. Allergy, **46**:310, 1981.
16. Hoffman, D.R., and Kozak, P.P.: Shared and specific allergens in mold extracts. J. Allergy Clin. Immunol., **63**:213, 1979.
17. Jones, H.E., et al.: Apparent cross-reactivity of airborne molds and the dermatophytic fungi. J. Allergy Clin. Immunol., **52**:346, 1973.
18. Kauffmann, H.F., and De Vries, K.: Antibodies against *Aspergillus fumigatus*. I. Standardization of the antigenic composition. Int. Arch. Allergy Appl. Immunol., **62**:252, 1980.
19. Kauffmann, H.F., and De Vries, K.: Antibodies against *Aspergillus fumigatus*. II. Identification and quantification by means of crossed immunoelectrophoresis. Int. Arch. Allergy Appl. Immunol., **62**:265, 1980.
20. Kim, S.J., et al.: Characterization of antigens from *Aspergillus fumigatus*. II. Fraction-

ation and electrophoretic, immunologic, and biologic activity. Am. Rev. Resp. Dis., **118**:553, 1978.
21. Kim, S.J., and Chaparas, S.D.: Characterization of antigens from *Aspergillus fumigatus*. I. Preparation of antigens from organisms grown in completely synthetic medium. Am. Rev. Resp. Dis., **118**:547, 1978.
22. Lefkovits, I., and Pernis, B.: Immunological Methods. Vols. I and II. New York, Academic Press, 1979, 1981.
23. Letunova, E.V., et al.: Purification and properties of beta-galactosidase of *Alternaria tenuis*. Biokhimiya, **46**:911, 1981.
24. Lowenstein, H.: Quantitative immunoelectrophoretic methods as a tool for the analysis and isolation of allergens. Prog. Allergy, **25**:1, 1978.
25. Malo, J.L., Paquin, R., and Longbottom, J.L.: Prevalence of precipitating antibodies to different extracts of *Aspergillus fumigatus* in a North American asthmatic population. Clin. Allergy, **11**:333, 1981.
26. Manning, M., and Mitchell, T.G.: Analysis of cytoplasmic antigens of the yeast and mycelial phases of *Candida albicans* by two-dimensional electrophoresis. Infect. Immunol., **30**:484, 1980.
27. Marsh, D.G., Meyers, D.A., and Bias, W.B.: The epidemiology and genetics of atopic allergy. N. Engl. J. Med., **305**:1551, 1981.
28. Merril, C.R., et al.: Ultrasensitive stain for proteins in polyacrylamide gels shows regional variation in cerebrospinal fluid patterns. Science, **211**:1437, 1981.
29. Murashige, T., and Skoog, F.: A revised medium for rapid growth and bioassays with tobacco tissue cultures. Phys. Plants, **15**:473, 1962.
30. Norman, P.S.: *In vivo* methods for the study of allergy. *In* Allergy: Principles and Practice. Edited by E. Middleton, C.E. Reed, and E.F. Ellis. St. Louis, C.V. Mosby, 1978.
31. Nyholm, L., Yunginger, J., and Lowenstein, H.: Immunochemical partial identity between ALT-1 and Ag-1 from *Alternaria alternata*. J. Allergy Clin. Immunol., **69**:97, 1981.
32. O'Farrell, P.H.: High resolution two-dimensional electrophoresis of proteins. J. Biol. Chem., **250**:4007, 1974.
33. Prince, H.E.: Mold fungi in the etiology of respiratory allergic diseases. III. Immunological studies with mold extracts. Skin tests with broth and washings from mold pellicles. Ann. Allergy, **2**:500, 1944.
34. Prince, H.E., Tatge, E.G., and Morrow, M.B.: Mold fungi in the etiology of respiratory allergic diseases. V. Further studies with mold extracts. Ann. Allergy, **5**:434, 1947.
35. Salvaggio, J., and Aukrust, L.: Mold-induced asthma. J. Allergy Clin. Immunol., **68**:327, 1981.
36. Sammons, D.W., Adams, L.D., and Nishizawa, E.E.: Ultrasensitive silver-based color staining of polypeptides in polyacrylamide gels. Electrophoresis, **2**:135, 1981.
37. Schaffer, N., Molomut, N., and Center, J.G.: Studies on allergenic extracts. I. A new method for the preparation of mold extracts using a synthetic medium. Ann. Allergy, **17**:380, 1959.
38. Schumacher, M.J., et al.: Primary interaction between antibody and components of *Alternaria*. II. Antibodies in sera from normal, allergic and immunoglobulin-deficient children. J. Allergy Clin. Immunol., **56**:54, 1975.
39. Schumacher, M.J., et al.: Primary interaction between antibody and components of *Alternaria*. J. Allergy Clin. Immunol., **56**:39, 1975.
40. Schumacher, M.J., and Jeffery, S.E.: Variability of *Alternaria alternata:* Biochemical and immunological characteristics of culture filtrates from seven isolates. J. Allergy Clin. Immunol., **58**:263, 1976.
41. Simmons, E.G.: Typification of *Alternaria, Stemphylium* and *Ulocladium*. Mycologia, **59**:67, 1967.
42. Slavin, R.G.: Allergic bronchopulmonary aspergillosis. *In* Allergy: Principles and Practice. Edited by E. Middleton, C.E. Reed, and E.F. Ellis. St. Louis, C.V. Mosby, 1978.
43. Solomon, W.R.: Aerobiology and inhalant allergens. I. Pollens and fungi. *In* Allergy: Principles and Practice. Edited by E. Middleton, C.E. Reed, and E.F. Ellis. St. Louis, C.V. Mosby, 1978.
44. Syverson, R.E., Buckley, H.R., and Campbell, C.C.: Cytoplasmic antigens unique to the mycelial or yeast phase of *Candida albicans*. Infect. Immunol., **12**:1184, 1975.

45. Vijay, H.M., et al.: Studies on *Alternaria* allergens. I. Isolation of allergens from *Alternaria tenuis* and *Alternaria solani*. Int. Arch. Allergy Appl. Immunol., **60**:229, 1979.
46. Weeke, B.: Crossed immunoelectrophoresis. Scand. J. Immunol. **2**(Suppl.1): 47, 1973.
47. Weir, D.M. (ed.): Handbook of Experimental Immunology. 3rd Ed. Oxford, Blackwell, 1978.
48. Wide, L., Bennich, H., and Johansson, S.G.O.: Diagnosis of allergy by an *in vitro* test for allergen antibodies. Lancet, **2**:1105, 1967.
49. Williams, C.A., and Chase, M.W. (eds.): Methods in Immunology and Immunochemistry, Vols. 1–5. New York, Academic Press, 1967–1977.
50. Yunginger, J.W., et al.: Studies on *Alternaria* allergens. III. Isolation of a major allergenic fraction (ALT-1). J. Allergy Clin. Immunol., **66**:138, 1980.
51. Yunginger, J.W., Roberts, G.D., and Gleich, G.J.: Studies on *Alternaria* allergens. I. Establishment of the radioallergosorbent test for measurement of *Alternaria* allergens. J. Allergy Clin. Immunol., **57**:293, 1976.
52. Zeiss, C.R., et al.: A solid phase radioimmunoassay for the quantitation of human reaginic antibody against ragweed antigen E. J. Immunol., **110**:414, 1973.

Chapter 7

DELAYED HYPERSENSITIVITY RESPONSES TO CUTANEOUS MYCOTIC AGENTS

Mervyn L. Elgart

Both cellular and humoral immunity appear to be active in host defenses against dermatophytic and *Candida* infections. Cell-mediated immunity is probably the most significant; this may be inferred from the increased incidence of *Candida* infections in patients with disturbed cellular immunity, as in individuals with lymphomas and reticuloses. Furthermore, this is seen in human transplant recipients, in whom cellular immunity has been purposely depressed by therapy, and in the aged, in whom diminished T-cell responsiveness has been accompanied by an increase in fungal infections (34). Even in atopic patients, in whom cell-mediated immune deficiency is not marked, chronic *Trichophyton rubrum* infections abound, and responses to therapy are poor (11). The dermatophytes almost always limit invasion to keratin rather than to living epidermis; therefore, the exact mechanisms involved in stopping a progressing infection have not been determined. Nevertheless, the presence of a functional cell-mediated immune system often marks the difference between an infection of a few days or weeks and a chronic, progressive, widespread disease.

Immunologic studies have been difficult to carry out in superficial fungal infections. Fungi are often inconstant in the production of antigenic substances, so that cultures derived from a single spore origin of a species may vary considerably in the production of antigenic materials. In addition, different growing conditions and media may affect antigen production. Media containing peptones, for example, may produce better growth of fungi, but the presence of nondialyzable material in the peptone may contribute to the nonspecific factors in antigenic extracts. Finally, experimental animals and humans may vary in their antibody response to these materials. The most consistent fungal cultures, "shake cultures," are not applicable to dermatophytes, which prefer surface growth (27). In general, type specificity is associated with the polysaccharide antigens, while broad species specificity is associated with the protein antigens.

DERMATOPHYTES

Trichophytin

Early investigators realized the importance of immunity in dermatophyte infections, and documented the appearance of delayed cutaneous hypersenitivity to a fungal product, "trichophytin," as a result of infection (3). Trichophytin, the product used for testing, was first produced by Neiser. Many different methods of preparing this antigen have been developed, and all have somehow involved the production of a water-soluble extract of hyphae in the culture medium. Invariably the product prompted many nonspecific reactants.

In 1960 Cruickshank et al. (5) published a method of purifying trichophytin. Fungal mycelia were separated from liquid culture medium and dried. The preparation was then extracted with various solvents, including water, ethylene glycol, and alcohol; infected animals were tested at each step for reactivity. The preparation method for the pure antigen may be outlined as follows. First, the acetone-powdered mycelium is extracted twice with ethylene glycol, and the combined extracts are filtered, dialyzed, and freeze-dried. This material is dissolved in water and precipitated with three volumes of ethanol for initial purification. The ethanol precipitate is redissolved in 1% sodium borate buffer, pH 8.5, and reprecipitated by the addition of 5% cetrimide solution. The pH of the solution is adjusted by the addition of 1 N NaOH until the precipitation is complete. This precipitate is then dissolved in 2 N acetic acid, and the solution poured into ethanol. The resulting precipitate is washed with a mixture of acetic acid-ethanol (50:50), ethanol, ether, and dried. To eliminate all traces of cetrimide, the precipitate should be dissolved in water and passed through a sulphonic acid resin in the hydrogen form, dialyzed, and freeze-dried (5).

Trichophytin powder was found to be a polysaccharide, with a molecular weight of 20,000 to 30,000 (5); it contained about 80% carbohydrate, of which glucose and mannose seemed to be present in equimolar concentrations. The polysaccharide component was attached to a protein or peptide component. Nozawa et al. (26) further studied trichophytin and determined that the protein component was necessary for delayed but not immediate reactions. Proteolytic digestion caused a loss of delayed responses, but did not affect the immediate urticarial responses.

Wood and Cruickshank (42) used this extract to test patients with active infections. They determined that about a third of patients with active infections showed a delayed reaction to the product, and that, conversely, the presence of delayed reactivity correlated well with infection. The reverse was not true, however, because many patients with infections did not react. Furthermore, the presence of an immediate wheal reaction was seen as often in noninfected as in infected individuals, but was particularly associated with atopy. They also reported a new kind of reactivity, the "retarded reaction," which occurred 5 to 7 days later and was eczematous. This reaction correlated with the tendency to develop "id" reactions. Other reports of this type of response have not been found.

Trichophyton mentagrophytes Infection

The immunologic events associated with dermatophyte infections in humans remained unclear for years. Although some investigators (3) believed that the mechanisms in human and animal ringworm were similar, others (6) concluded that there was little acquired immunity in humans. Desai (6) did mention, however, that infections in children lasted longer than in adults, which probably was due to the children's lack of previous exposure to the fungal agent. In this situation the spreading and inflammatory phases probably last longer, until significant immunologic response can be generated.

Jones et al. (17) selected and tested dermatophyte-free individuals (fungal virgins) from a prison population. Subjects were chosen on the basis of a negative physical appearance of infection, negative history of dermatophyte infection, and lack of response to trichophytin skin testing. Even patients with a minimal immune response were eliminated, because such individuals frequently might have an anamnestic response, and thus show more reactivity after testing.

It was found that an infection to *Trichophyton mentagrophytes* could be established in each subject with a minimal number of spores; six spores was the smallest number tested. The possibility of infection was greater when the number of spores was increased; greater inocula were associated with greater primary irritant reactions, which were seen initially. For the first few days the infection spread in a peripheral fashion. After about 10 to 17 days the lesions became intensely inflammatory (Fig. 7–1). This inflammatory change correlated with a change in trichophytin reactivity from negative to positive. The size of the "ring" was limited to the size that had appeared at the time of the subject's conversion. Peripheral spreading ceased, the lesion gradually became culturally negative, and the inflammation subsided. Scaling or pigment change persisted for 2 weeks.

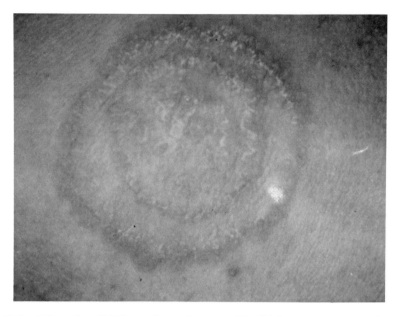

Fig. 7–1.   Fifteen-day-old inflammatory lesion caused by *Trichophyton mentagrophytes.*

The situation in reinfection was somewhat different. An intense inflammation was noted after occlusion of 96 hours. The investigators tested with dead hyphae, and showed that the latter inflammation was actually an allergic contact dermatitis (17). This appeared significantly more acute, and longer lasting, than the irritant reaction that appeared when fungi were placed on the skin of previously unexposed individuals. In the latter case the area of infection never enlarged, and it was impossible to demonstrate the fungal agent after the tenth day. The remaining inflammation gradually subsided. In the initial reactions in unexposed individuals the size of the infection was limited by the speed at which delayed hypersensitivity developed. Similarly, the presence of delayed hypersensitivity was associated with severe inflammation and with healing of the primary lesion.

The course of a dermatophyte infection must therefore be seen in dynamic rather than static terms. There are four stages in dermatophyte infection in experimental animals and humans; incubation, spreading, climax or inflammation, and clearing (17). The clearing phase is usually associated with an enhanced protection against subsequent infection. Furthermore, this protection is not species-specific but is usually effective against many species of dermatophytes.

Other *Trichophyton* and *Microsporum* Infections

Trichophyton mentagrophytes infections have been used almost exclusively in this field. However, Hall (9) has reported on *T. verrucosum* infections, and suggested that the failure to observe recurrent infections among farmhands must be due to immunity. Similarly, the presence of positive delayed hypersensitivity has been investigated in tinea capitis infection with *T. tonsurans* among children (29). There are basically two distinct types of *T. tonsurans* infections, the kerion type (Fig. 7–2), or inflammatory lesion, and the black dot type, in which inflammation is minimal (Fig. 7–3). In the kerion type the presence of the inflammation is correlated with the activation of delayed hypersensitivity, while in the black dot (noninflammatory) type delayed hypersensitivity is inactive. Allen et al. (1) have reported a case in which *Microsporum audouini* infection became generalized. In this patient anergy and defective lymphocyte transformation were demonstrated. Treatment with plasma restored some of the lymphocyte blastogenesis; clearing of the infection was then achieved.

Many infections with *T. rubrum* are chronic, and appear in atopic individuals. Hanifin et al. (10) showed that these individuals had immediate but not delayed hypersensitivity to trichophytin, while patients with *T. mentagrophytes* infection almost always responded with delayed hypersensitivity. Similar responses were shown by Hay and Brostoff (12) and by Kaaman (19). Sloper (36) found that experimental infections with *T. rubrum* healed spontaneously in normal (nonatopic) individuals; therefore, the diminished ability of the atopic individual to produce delayed hypersensitivity responses may imply chronic infections, not a peculiar pathogenicity of a particular organism.

The exact nature of the disturbance in the atopic patient is unclear. Originally patients with atopic dermatitis were believed to be deficient in their ability to show delayed hypersensitivity. Hunziker and Brun (15) may have demonstrated an alternative mechanism: the antigen may be neutralized by serum antibodies during the immediate reaction. When a second dose was injected into the same site 20 minutes after the initial dose, a strong delayed reaction was indeed elicited, demonstrating that these individuals do not have a totally depressed cell-mediated immunity. Similarly, Jones et al. (16) have described atopic patients who were able to synthesize an immediate-reacting antibody that reacted with fungi and prevented involvement of cell-mediated antibody. On the other hand, in patients with a low incidence of positive delayed reactions to trichophytin, many were found to have significant reduction in all tests of cell-mediated immunity (38).

Lymphocyte Tests

Most reported lymphocyte tests have used the classic skin test and the routine 48-hour reaction. Normal cellular immunity, however, appears to require a number of processes, which may be individually separated and tested. These include lymphocyte blastogenesis, lymphokine

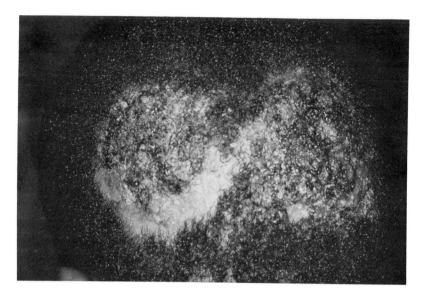

Fig. 7–2. Tinea capitis infection showing a "kerion" type caused by *Trichophyton tonsurans.*

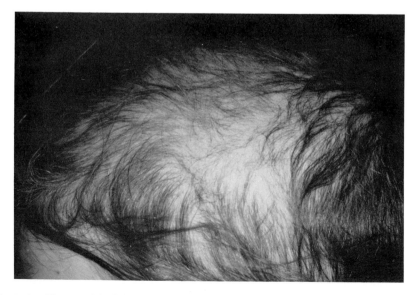

Fig. 7–3. Tinea capitis showing a noninflammatory "black dot"-type infection caused by *Trichophyton tonsurans.*

production, and monocyte chemotactic responsiveness. Chronic fungal disease may result when any of these three processes is missing.

Hay and Brostoff (12) studied lymphocyte migration and found no difference between atopic patients and those who had just recovered from a dermatophyte infection. Hanifin et al. (10) compared patients having *Trichophyton mentagrophytes* infections and positive delayed hypersensitivity with patients having chronic *T. rubrum* and positive immediate hypersensitivity. The lymphocyte transformation test in response to trichophytin antigen showed positive results in *T. mentagrophytes* patients and negative results in *T. rubrum* patients and in noninfected controls. Hunziker and Brun (15) have reported on lymphocyte transformation in atopic patients having negative delayed tests, including three patients with elevated IgE levels. The trichophytin antigen stimulation was increased in only one patient. Kaaman et al. (20) found that patients with acute infections and positive trichophytin tests showed positive lymphocyte stimulation, while only 50% of patients with negative delayed tests and/or positive immediate tests showed lymphocyte stimulation. The finding of positive lymphocyte stimulation in some patients with chronic infections indicates some degree of cellular immunity in such patients. It is not known why there is negative lymphocyte stimulation in some patients. It has been suggested, however, that blocking antibodies are active in negative tests.

Walters et al. (40) have used the leukocyte adherence inhibition test. They reported these tests to be positive and species-specific in dermatophyte infections; therefore, patients who reacted to *Trichophyton mentagrophytes* did not react to *T. rubrum* or to *Epidermaphyton floccosum*. A contrasting previous report had shown cross reactions for all fungal agents tested (10).

Mechanism of Fungus Inhibition

The exact mechanism by which fungi are inhibited or destroyed in an immune-competent individual has not been clearly demonstrated. Jones et al. (17) have suggested that the cellular reactivity to infection produces inflammation that allows the stratum corneum to be penetrated by a normal serum component. That component, an antifungal factor, was first suspected when Lorincz et al. (23) were unable to produce infection in mice by injecting dermatophytes abdominally. The dermatophytes remained viable and were recovered. A serum factor, inhibitory to fungi, was suggested; it was later found to be present in normal serum, both from previously infected or noninfected individuals (4). This factor was subsequently shown to be an unsaturated transferrin (21).

Berk et al. (2) studied five patients with annular tinea corporis using tritiated thymidine labeling. Skin turnover time in the epidermal cells adjacent to the infection site was higher than that of normal skin but the skin within the lesion had a normal labeling. The increased skin turnover time may be important in restricting invasion of the stratum corneum by the dermatophyte, or this change might modify the epidermal barrier to allow penetration of the serum inhibitory factors to the skin surface.

An alternative hypothesis to the Berk et al. findings may be direct destruction of fungi by activated T lymphocytes (thymus-dependent cells), possibly by secretion of lymphokines that diffuse to the surface. The necessity for an intact cell-mediated immunity is again required.

Vaccination

Jones et al. (17) considered a vaccine to dermatophytes for protection against further infection. The resistance achieved would not be permanent but might be helpful, as they concluded, for persons at increased risk, such as soldiers, athletes, or those living in the tropics for prolonged periods. Live vaccine could not be used because of the possibility of infection in those who had inadequate resistance, such as the atopics with a chronic infection. Theoretically, at least, an immunized person, like previously exposed individuals, might require a much larger inoculum for disease to be produced.

In summary, activation of cellular immunity appears to occur during the course of a dermatophyte infection in a previously noninfected individual. The appearance of that immunity correlates well, although not invariably, with the development of inflammation in these infections. The presence of immunity can be measured by the delayed trichophytin test and by lymphocyte blastogenesis.

Infection in a previously exposed person, in whom cellular immunity is active, is usually brief. Inflammation occurs early, and the infection is abated. The mechansim by which the infection is stopped may be the establishment of unsaturated transferrin in the stratum corneum, resulting in inhibition of the fungi. An alternate explanation is that fungi are directly destroyed by activated thymus-dependent lymphocytes or by lymphokines produced by them.

Infection in an atopic individual produces an IgE elevation and an immediate response to trichophytin skin testing. There is either depression of delayed hypersensitivity or blocking of the antigen by IgE, promoting persistent infections in such patients.

CANDIDOSIS

A discussion of cellular responses in candidosis must take into account the widespread colonization by *Candida albicans* (or other species of *Candida*) without production of disease, and the events that occur in limited cutaneous infections, chronic mucocutaneous candidosis, and disseminated candidosis or *Candida* septicemia. Investigation of cellular responses in these areas has been difficult, and not entirely satisfactory.

Although it is believed that cellular responses are intimately concerned with the depth of infection, confirming work to support this statement is lacking.

In candidosis, the conversion from the yeast phase to the mycelial phase (or pseudomycelia) often occurs with the production of actual infection. The presence of hyphae, rather than of budding yeast cells only, has been used to distinguish between colonization (yeast) and infection (hyphae mixed with yeasts). Manning and Mitchell (25) examined yeast and mycelial phase antigens in strains of *Candida*. They demonstrated hundreds of different protein fractions, but no single antigen determinant seemed to be exclusive to either the yeast or mycelial phase.

Superficial Clinical Disease and Immunologic Response

Clinical disease in candidosis was investigated by Maibach and Kligman in a classic paper published in 1962 (24). They applied *Candida* organisms to the skin and found that moisture and occlusion were necessary for the production of disease. Application without occlusion did not produce disease in any area except under the foreskin. With occlusion, clinical disease was first apparent at 36 to 72 hours as erythema, erosions, and pustules mimicking the natural infection (Fig. 7–4). Microscopically, lymphocytic infiltrate was present (in 24 hours) before clinical disease was detected. Later a subcorneal pustule was noted. In the roof of the pustule small numbers of organisms were observed, both in budding yeast and in mycelial forms. Organisms were never observed to invade the living epidermis. This experimental procedure produced infection in every patient inoculated with 1.5×10^6 organisms. The threshold of infection was 10^6 organisms, and the 50% infection rate was 6×10^5 organisms.

Oidiomycin is a *Candida* extract antigen that has been used for skin testing in a method similar to that used with trichophytin. Maibach and Kligman (24) prepared the antigen from a sterilized filtrate of a maltose broth culture of *C. albicans*. The antigen produced a positive reaction in 100% of the 20 patients in whom it was tested, usually within 24 hours. When the antigen was diluted 1:100 only 6 of 20 patients reacted. Investigators attempted to see how the reaction was modified after infection by reinfecting individuals or by keeping them continuously infected for 5 consecutive weeks. Reactions to this material did not change after repeated infections. There are no reports in the literature regarding *Candida* testing in nonexposed individuals to be compared to the studies of Jones et al. (17) with dermatophytes.

When *Candida* cells were killed and desiccated, and the cell fragments applied to the skin under occlusion, an inflammation that mimicked the experimental infection appeared in every case (24). Maibach and Kligman concluded that the yeast infection as seen clinically on the skin is a

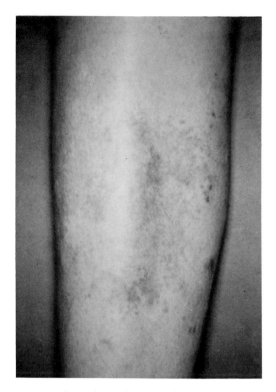

Fig. 7–4.   Erythematous papules and pustules of candidosis in an unusual location following plastic wrap occlusion of the leg.

biologic contact dermatitis of the primary irritant type. Furthermore, because these irritants are bound to the cell walls or to cell substance, and are not present in the water-soluble material produced in culture, these investigators believed that it was the killed cell rather than the infection that produced the clinical disease. Similar results were reported by Rebora et al., in which the presence of a serum inhibitory factor was disputed (33).

Results of experimental work in rodents by Ray and Wuepper (32) showed that infections could easily be produced by epicutaneous application of *Candida albicans* or *C. stellatoidea* under occlusion for 24 hours. The pustular response within the epidermis was again demonstrated, and was found to be caused by either live or killed organisms. The presence of neutrophils was found to be due to the activation of the alternative (properdin) pathway of complement, which was shown to occur with viable organisms, killed organisms, soluble cell wall polysaccharides, and mannans (30,31). The subsequent release of chemotactically active complement fragments C3a, C5a, and C567 mediated the migration of neutrophils (41). Furthermore, unlike what was reported previously (24), the organisms were found to invade the stratum malpighii as well as the stratum corneum. In the case of *C. tropicalis*, for example, although it would not invade the stratum corneum it could grow in the stratum malpighii if the stratum corneum were removed (13).

Ray and Wuepper concluded that host defenses against *Candida* invasion included an intact stratum corneum, a functioning complement system, and an adequate neutrophilic response (30–32). In patients in whom the surface epithelium is irritated by catheters or by poorly fitting dentures, the organisms grow and invade the surrounding tissues. In mice with complement deficiencies, especially C5 deficiencies, inflammatory response to *Candida* was minimal and proliferation of mycelia was deep. In cases of defective neutrophil chemotaxis, such as the hypergammaglobulinemia syndrome (hyper-IgE syndrome), increased *Candida* infections resulted.

Chronic Mucocutaneous Candidosis

Chronic mucocutaneous candidosis (Fig. 7–5) can be seen in two situations: those associated with basic immunopathologic disorders, and those in which there is no other pathology and the candidosis is the primary infection. In the basic disease category, patients with Swiss-type agammaglobulinemia, hereditary thymic dysplasia, and the DiGeorge syndrome may have chronic *Candida* infections (13). Therapy with oral or systemic anti-*Candida* drugs has been helpful, but patients usually succumb to the basic disease.

Cases of chronic mucocutaneous disease are a mixed group. These patients may include familial chronic mucocutaneous candidosis, disseminated mucocutaneous candidosis, the *Candida*-endocrinopathy syndrome, and chronic mucocutaneous candidosis occurring later in life, as with denture stomatitis, or associated with occult malignancy (13). Most of these patients have depressed cell-mediated immunity, especially the familial and disseminated types. Iron deficiency was found, particularly in the familial, endocrine, and late onset types.

Several interesting treatments have been used, suggesting explanations of the basic immunologic process. Iron therapy was reported to be beneficial in 9 of 11 patients (34). In these patients iron was given in addition to the usual topical antifungal preparations. The reason for the positive response to iron was not clear, although some of the patients were thought to have an impaired iron absorption. An unsaturated transferrin is found in normal serum and seems to be important in resistance to yeast infections. This could be explained by the presence of the "antiyeast serum factor," which was mentioned earlier. Unsaturated transferrin, by chelating iron necessary for yeast growth, inhibits *Candida in vitro*. This process is reversed when iron is added to the medium. Per-

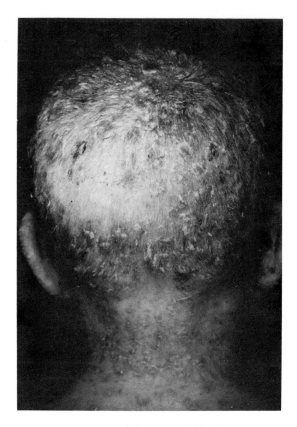

Fig. 7–5. Chronic mucocutaneous candidosis in a child with no other immune deficiency.

haps oral or parenteral iron helps to reconstitute this molecule. Patients who respond to iron not only improve clinically but also develop positive delayed hypersensitivity responses and positive lymphocyte function tests (34).

Another point of interest in these patients is "transfer" factor, a substance elaborated by lymphocytes in response to a particular antigen (22). This factor can be transferred to a nonexposed patient, who will react as if his own cells have been in contact with the particular infectious agent. The substance appears to be produced by lymphocytes of the thymus-dependent group. It is a dialyzable substance of less than 10,000 daltons, and is neither an immunogenic nor an immunoglobulin fragment (22).

When transfer factor from normal individuals was used, delayed skin test sensitivity to *Candida* could be produced in those who did not previously have positive reactions. Furthermore, migration inhibition factor (MIF) could then be produced by the recipients. This was only successful in patients who were unable to synthesize their own transfer factor. Patients who could produce their own factor, but whose MIF was inhibited, were not helped (14).

Cimetidine is another treatment that has shown promise and that may help to elucidate the pathophysiology of the disorder. Histamine in guinea pigs has been shown to depress cell-mediated immunity as measured by reduced skin test reactivity, reduced antigen-induced lymphokine migration inhibitory factor production, and reduced lymphocyte proliferation response (35). The histamine-induced suppression was completely blocked by treatment with H-2 blocking drugs. Peptic ulcer patients who receive a course of cimetidine therapy have been shown to have enhanced cell-mediated responses to skin test antigens (18). In four patients with the familial type of chronic mucocutaneous candidosis, a 4-week course of cimetidine produced dramatic changes in cellular reactivity. Skin tests became strongly positive in four patients, and two patients produced MIF that had not previously been present. Lymphocyte transformation tests were not altered by therapy; changes reverted to pretreatment level 4 weeks after discontinuing treatment. The drug was not given long enough to assess clinical improvement.

Lymphocyte Tests

Most patients have normal cellular immune responses, as measured by lymphocyte transformation studies. Sohnle and Collins-Lech (37) found that 44 of 49 normal individuals had positive responses to *Candida*, while Elliott and Hanifin (7) found positive responses in 25 out of 27. Atopic patients were found to have fewer positive responses (16 out of 25). The relationship of these tests to active or recent disease is not clear. Perhaps the high percentage of positive reactions is due to continuous *Candida* colonization in most individuals.

Piccolella et al. (28) have described a purified polysaccharide fraction, Mangion purified polysaccharide (MMPS), obtained from autoclaved *Candida albicans* yeast cells after precipitating the supernatant fraction with 95% ethanol and then deproteinizing the preparation with a mixture of butanol and chloroform. MMPS is a polysaccharide containing glucose and mannose, which contains 3.1% protein. Purified human T or B cells were found not to respond to MMPS. When a mixture of these two types of cells was used, however, the mitogenic response occurred. Both the helper effect of T cells on B cells and the helper effect of B cells on T cells were noted. The T-cell helper effect may occur in part through soluble mediators. The helper effect did not require blastogenesis on the part of the helper cell.

Stobo et al. (39) examined four patients with disseminated *Candida* infection. All manifested *in vitro* T-cell hyporeactivity. This hyporesponsiveness was traced to the action of suppressor T cells rather than to an inherent decreased responsiveness. Furthermore, the lymphocytes could be restimulated to react to the appropriate antigen after 7 days of culture without mitogenic or antigenic stimuli.

Role of Bacterial and Topical Steroids

Candida albicans has been found to be inhibited by the presence of gram-negative bacilli, an observation consistent with the increase in *Candida* infections seen in patients taking broad-spectrum antibiotics. Perhaps bacteria and *Candida* compete for nutrients. Normal non-sterile saliva can support the growth of *Candida*, up to 10^3 organisms per ml. The number or organisms may be increased if glucose is added to the saliva. In previously sterilized saliva, however, *Candida* grew to a concentration of 10^6 cells ml. In experimental infections, Rebora et al. (33) found that, to infect plaques of atopic dermatitis or psoriasis by the usual techniques of epicutaneous application of organisms and occlusion, the tested skin first had to be sterilized with alcohol.

The use of topical steroids or short-term systemic steroids does not affect experimental *Candida* infections (24). Long-term steroids may do so by depressing cellular immunity or by enhancing glucose concentrations in tissue and possible diabetic conditions. In granuloma gluteale infantum, the use of topical steroids sometimes appears to enhance the local *Candida* infection and to produce a granulomatous response (34).

In summary, then, most patients respond to superficial yeast infections by the activation of complement and by neutrophil chemotaxis. Conditions that change the epidermal barrier, such as maceration, changes in glucose concentration or in the ecology, and topical steroids may all be active in superficial yeast infections.

Cell-mediated immunity can be demonstrated in most normal adults, but only in some patients with chronic mucocutaneous candidosis. Restoration of cell-mediated immunity is associated with improvement in the clinical course of chronic mucocutaneous candidosis. The presence of a functional cellular immune system, a functioning complement system, and neutrophil chemoresponsiveness is necessary in limiting *Candida* infections and in preventing the development of chronic mucocutaneous candidosis. An adequate level of transferrin seems to be active in resistance to *Candida*, but its role is unclear.

PITYRIASIS VERSICOLOR

Pityriasis versicolor (formerly called tinea versicolor) is a frequent chronic recurrent infection on the chest, neck, and back. The causative organism has been called *Pityrosporum furfur* (formerly known as *Malassezia furfur*). The organism can be cultured by overlaying the medium with olive oil, because this is a lipophilic organism. The disease has been reproduced in humans and rabbits by inoculation (8).

Sohnle and Collins-Lech (37) have studied the responses of patients with pityriasis versicolor and normal controls to test for cell-mediated immunity. *Pityrosporum furfur* extract was able to provoke lymphocyte transformation in both patients and controls. The patients, however, produced less migration inhibition factor than the controls when stimulated by the *P. furfur* extracts. There may be, therefore, a group of predisposed individuals who normally may be easily infected by this organism. Other more resistant individuals may become infected when a warmer climate causes improved growing conditions and overcomes the normal host defenses.

REFERENCES

1. Allen, D.E., et al.: Generalized *Microsporum audouini* infection and depressed cellular immunity associated with a missing plasma factor required for lymphocyte blastogenesis. Am. J. Med., **63**:991, 1977.
2. Berk, S.H., Penneys, N.S., and Weinstein, G.D.: Epidermal activity in annular dermatophytosis. Arch. Dermatol., **112**:485, 1976.
3. Bloch, B., and Massini, R.: Studien über immunitat und über empfindlichkeit bei hyphomyzeternerkrankungen. Z. Hyg. Infekt. Krankh., **36**:68, 1909.
4. Carlisle, D.H., et al.: Significance of serum fungal inhibitory factor in dermatophytosis. J. Invest. Dermatol., **63**:239, 1974.
5. Cruickshank, C.N.D., Trotter, M.D., and Wood, W.R.: Studies on trichophytin sensitivity. J. Invest. Dermatol., **35**:219, 1960.
6. Desai, S.C.: Immunology of dermatophytosis. *In* Proceedings of the Intern. Cong. on Dermatol. Edited by D. Pillsbury and C.S. Livingood. Amsterdam, Excerpta Medica Foundation, 1962.
7. Elliot, S.T., and Hanifin, J.M.: Delayed cutaneous hypersensitivity and lymphocyte transformation. Arch. Dermatol., **115**:36, 1979.
8. Faergemann, J.: Tinea versicolor and *Pityrosporum orbiculare*. Acta Derm. Vener. [Suppl.] (Stockh.), **86**:1, 1979.
9. Hall, F.R.: Ringworm contracted from cattle in Western New York state. Arch. Dermatol., **94**:35, 1966.
10. Hanifin, J.M., Ray, L.F., and Lobitz, W.C.: Immunological reactivity in dermatophytosis. Br. J. Dermatol., **90**:1, 1974.
11. Hay, R.J.: Failure of treatment in chronic dermatophyte infections. Postgrad. Med. J., **55**:608, 1979.
12. Hay, R.J., and Brostoff, J.: Immune responses in patients with chronic *Trichophyton rubrum* infections. Clin. Exp. Dermatol., **2**:373, 1977.
13. Higgs, J.M., and Wells, R.S.: Chronic muco-cutaneous candidosis: New approaches to treatment. Br. J. Dermatol., **89**:179, 1973.
14. Horsmanheimo, M., et al.: Immunologic features of chronic granulomatous mucocutaneous candidiasis before and after treatment with transfer factor. Arch. Dermatol., **115**:180, 1979.
15. Hunziker, N., and Brun, R.: Lack of delayed reaction in presence of cell-mediated immunity in trichophytin hypersensitivity. Arch. Dermatol., **116**:1266, 1980.
16. Jones, H.E., Reinhardt, J.H., and Rinaldi, M.G.: A clinical, mycological, and immunological survey for dermatophytosis. Arch. Dermatol., **108**:61, 1973.
17. Jones, H.E., Reinhardt, J.H., and Rinaldi, M.G.: Acquired immunology to dermatophytes. Arch. Dermatol., **109**:840, 1974.
18. Jorizzo, J.L., et al.: Cimetidine as an immunomodulator: Chronic mucocutaneous candidiasis as a model. Ann. Intern. Med., **92**:192, 1980.
19. Kaaman, T.: The clinical significance of cutaneous reactions to trichophytin in dermatophytosis. Acta Dermatol. Vener. (Stockh.), **58**:139, 1978.
20. Kaaman, T., Petrini, B., and Wasserman, J.: In vivo and in vitro immune responses to trichophytin in dermatophytosis. Acta Dermatol. Vener. (Stockh.), **59**:229, 1979.
21. King, R.D., et al.: Transferrin, iron and dermatophytes, I. Serum dermatophyte in-

hibitory component definitively identified as unsaturated transferrin. J. Lab. Clin. Med., **86**:204, 1975.

22. Lawrence, H.S.: Transfer factor and cellular immune deficiency disease. N. Engl. J. Med., **283**:411, 1970.
23. Lorincz, A.L., Priestly, J.D., and Jacobs, P.H.: Evidence for a humoral mechanism which prevents the growth of dermatophytes. J. Invest. Dermatol., **31**:15, 1958.
24. Maibach, H.I., and Kligman, A.M.: The biology of experimental human cutaneous moniliasis. Arch. Dermatol., **85**:113, 1962.
25. Manning, M., and MItchell, T.G.: Analysis of cytoplasmic antigens of the yeast and mycelial phases of *Candida albicans* by two-dimensional electrophoresis. Infect. Immunol., **30**:484, 1980.
26. Nozawa, Y., et al.: Immunochemical studies on *Trichophyton mentagrophytes*. Sabouraudia, **9**:129, 1971.
27. Pepys, J., and Longbottom, J.L.: Immunological methods in mycology. *In* Handbook of Experimental Immunology. 3rd Ed. Edited by D.M. Weir. Oxford, Blackwell, 1978.
28. Piccolella, E., Lombardi, G., and Morelli, R.: Human lymphocyte-activating properties of a purified polysaccharide from *Candida albicans*. J. Immunol., **125**:2082, 1980.
29. Rasmussen, J., and Ahmed, A.R.: *Trichophyton* reactions in children with tinea capitis. Arch. Dermatol., **114**:371, 1978.
30. Ray, T.L., et al.: Purification of a polymannan from *Candida albicans* which activates serum complement. Clin. Res., **25**:100A, 1977.
31. Ray, T.L., and Wuepper, K.D.: Activation of the alternative (properidin) pathway of complement by *Candida albicans* and related species. J. Invest. Dermatol., **67**:700, 1976.
32. Ray, T.L., and Wuepper, K.D.: Experimental cutaneous candidiasis in rodents. Arch. Dermatol., **114**:539, 1978.
33. Rebora, A., Marples, R.R., and Kligman, A.M.: Experimental infection with *Candida albicans*. Arch. Dermatol., **108**:69, 1973.
34. Roberts, S.O.B., and Mackenzie, D.W.R.: Mycology. *In* Textbook of Dermatology. 3rd Ed. Vol. I. Edited by A. Rook, D.S. Wilkinson, and F.J.G. Ebling. Oxford, Blackwell, 1979.
35. Rocklin, R.E.: Modulation of cellular immune responses in vivo and in vitro by histamine receptor-bearing lymphocytes. J. Clin. Invest., **57**:1051, 1976.
36. Sloper, J.C.: A study of experimental human infections due to *Trichophyton rubrum*, *Trichophyton mentagrophytes*, and *Epidermaphytin floccosum*, with particular reference to the self-limitation of the resultant lesions. J. Invest. Dermatol., **25**:21, 1955.
37. Sohnle, P.G., and Collins-Lech, C.: Relative antigenicity of *P. orbiculare* and *C. albicans*. J. Invest. Dermatol., **75**:279, 1980.
38. Sorenson, G.W., and Jones, H.E.: Immediate and delayed hypersensitivity in chronic dermatophytosis. Arch. Dermatol., **112**:40, 1976.
39. Stobo, J.D., et al.: Suppressor thymus-derived lymphocytes in fungal infection. Clin. Invest., **57**:319, 1976.
40. Walters, B.A.J., Beardmore, G.L.. and Halliday, W.J.: Specific cell-mediated immunity in the laboratory diagnosis of dermatophytic infections. Br. J. Dermatol., **94**:55, 1976.
41. Ward, P.A., et al.: The role of serum complement in chemotaxis of leucocytes in vitro. J. Exp. Med., **122**:327, 1965.
42. Wood, S.R., and Cruickshank, C.N.D.: The relation between trichophytin sensitivity and fungal infection. Br. J. Dermatol., **74**:329, 1962.

Chapter 8

IMMEDIATE HYPERSENSITIVITY RESPONSES TO FUNGAL AGENTS

Kjell Aas

Lars Aukrust

This chapter is restricted to immediate hypersensitivity (allergic) responses on the molecular, cellular, and tissue levels. Understanding these responses is necessary to comprehend the clinical manifestations of allergy to fungal agents and to develop rational therapeutic approaches. Clinical aspects are referred to only briefly, in relation to relevant investigations of respiratory allergy.

COMPONENTS OF IMMEDIATE ALLERGIC RESPONSES

The development and manifestations of allergies to fungal agents seem to follow the same laws of nature and biology as other allergies to environmental agents. Allergy depends on immunologic sensitization; sensitization within the immunoglobulin E system (IgE) that produces reaginic (homocytotropic) antibodies is most important for immediate hypersensitivity, although another kind of homocytotropic antibody belonging to the immunoglobulin G system (IgG) may also play a minor part. Immediate reactions may also be caused by antigen-IgG immunocomplexing, resulting in activation of the complement cascade. This possibility, however, will not be discussed further here.

The characteristics of the molecule, immunogen, that causes the primary sensitization in the IgE system is not known. It is likely that the determinants causing sensitization, immunogenic determinants, are identical to, or at least rather similar to, the antigenic determinants that

133

elicit the allergic responses later on. In this text the latter are called allergenic determinants, the molecules carrying them are called allergens, and the source substance for allergens is called the allergen source.

In the IgE system several host factors appear important for sensitization, such as genetic disposition, allergen exposure, route of entry of allergens, local microcirculation, and homeostatic factors. A set of genetically determined dispositions may be decisive (16). To make the genetic disposition manifest, however, sufficient exposure to agents with appropriate antigenic potential is essential.

Mould spores may outnumber pollen grains by a ratio of several thousand to one, even for one species. Clinical experience has demonstrated that the abundance of environmental fungal agents represents sufficient exposure for a large number of individuals to become sensitized. Since the time of Blackley (10) it has been known that the inhalation of airborne mould spores may produce respiratory allergic symptoms.

The spores as such, and allergenic material extracted from them into the secretions, are likely to interact with specifically sensitized mast cells found in the respiratory passages and tissues. Due to the small spore size, moulds are more likely to sensitize and give allergic reactions in the lower respiratory passages than in the nasal tissues. The actual incidence of rhinitis and asthma induced by inhaled fungal spores is not certain, though, and may differ with geographic location, topography, and climate.

Some fungi may also be cultured from normal human sputum, especially from individuals with immunodeficiencies or from those on long-term treatment with antibiotics or local respiratory corticosteroids. The importance of allergic reactions to such saprophytic fungi has not been established.

Fungi are also used in the production of different kinds of food, or they may be contaminants in fruit and vegetables. There is no convincing evidence that ingested fungi are important offenders in allergic diseases, except in rare cases. Why this is so is not clear and, to our knowledge, this has not been studied. It could be that major inhalant allergens in fungi are labile and become easily denatured by processing and digestion. If this is true, it would be similar to the situation seen in allergies to wheat, because wheat allergens are labile and easily destroyed. Thus, inhalant allergy to wheat is found among many bakers, although the patients can eat wheat without experiencing any untoward reaction.

The Frontline Encounter

When airborne mould particles are inhaled they encounter a number of defense barriers such as mucus, which contains antibodies and enzymes. When spores are arrested in the air passages the wet environment facilitates immediate extraction of antigenic material.

The epithelial cells lining the bronchial mucosa are tightly bound to-

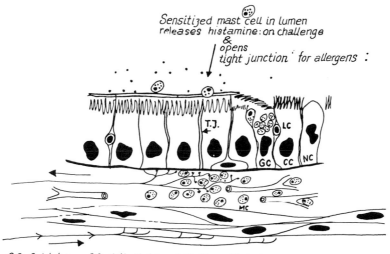

GC=Goblet c   CC=Ciliated c   NC=Non-ciliated c
T.J.=Tight junction   LC= Lymph c   MC=Mast cell

Fig. 8–1. Important components of the air passages in regard to the introduction of mould immunogens and allergens.

gether but the continuity may be broken at so-called tight epithelial junctions (14). These junctions may open up to allow entry of antigenic macromolecules into the deeper layers. Tight epithelial junctions may be opened during inflammatory or edematous reactions in the mucosa, possibly due to irritants, fumes, and chemical substances inhaled or released in the bronchial lumen. There are a few intraluminal mast cells that may release mediator substances upon appropriate allergen challenge. This opens up the tight junctions, and may allow more allergens into the deeper tissues (Fig. 8–1). Toxins produced by fungi (see below, "Toxicity") may have direct tissue effects when released into bronchial secretions and may enhance the sensitization and subsequent responses to exposure of sensitized tissues.

Allergens for the IgE system have been shown to be proteins, glycoproteins, and peptides or haptenic material bound to proteins or peptides (1). The allergens that have been characterized so far share a few molecular characteristics. Any antigenic protein molecule may act as an allergen in an individual who has a genetic disposition for IgE antibody responsiveness under given circumstances. Results of recent studies have shown, however, that exactly the same molecules that most commonly act as allergens in humans also act as allergens in mice and rats in regard to IgE antibodies (17). If confirmed, this would suggest that there may be certain molecular characteristics that preferentially sensitize within the IgE system.

The genetic mechanisms in question are far from being elucidated.

Evidence has been presented for a genetic disposition in rather general terms for the development of IgE-mediated allergy. Sensitization with minimal exposure, at least to certain minor allergens, seems to be genetically determined with respect to specificity (16).

Many of the most common allergens may belong to so-called universal antigens, which means that all or most humans are thought to possess the immunogenetic potential to react to them. IgE-mediated allergy to such allergens would then depend only on the type, degree, and conditions of allergen exposure. Because only a small proportion of sensitive individuals are allergic to fungi in spite of their overall abundance, it is likely that development of IgE-mediated allergy to fungal antigens depends to some extent on antigen-specific genetic dispositions.

Allergens are proteins or glycoproteins of small to medium size (2,000 to 100,000 daltons). As such they probably are processed by macrophages before being presented to the complex system of B and T immunocytes (Fig. 8–2).

IgM, IgA, IgG, IgD, or IgE antibodies may then be produced locally or within the system, depending on the site of introduction. Lymphoid tissue is found throughout the respiratory passages, from the nasopharynx to the respiratory bronchioles and alveolar ducts. Bronchial lymphoid nodules may be seen as bulging islands between ciliated and nonciliated epithelial cells. They are separated from the inspired air only by a single layer of flat nonciliated epithelium-containing lymphocytes (Fig. 8–3).

Lymphocytes are found throughout the lung tissues in such nodules

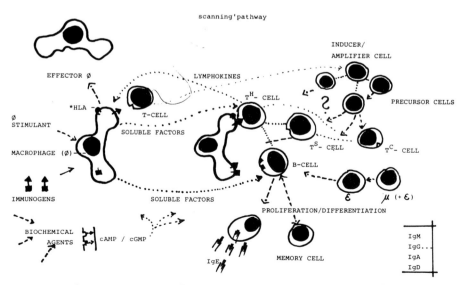

Fig. 8–2. The immunogens and allergens are processed by macrophages and presented to a complex set of immunocompetent cells.

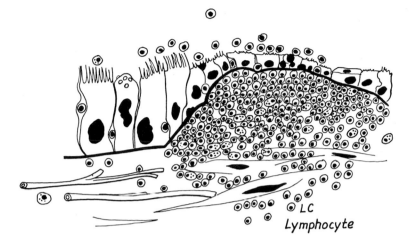

Lymphocyte

Fig. 8–3. Lymphoepithelial nodules may be seen as bulging islands between ciliated and nonciliated epithelial cells in the bronchial mucosa.

but are also found in cell aggregates, in more diffuse collections, and as single cells. Among them thymus-dependent lymphocytes (T cells) constitute a very mobile immunocompetent population. They migrate from the blood vessels into submucosal tissues of the respiratory passages. They may remain in the submucosa for some time before they enter the blood vessels again to recirculate. The "bursa equivalent" or bone marrow-derived B cells migrate slowly within lymphatic tissues. When stimulated by the appropriate antigen, they transform into antibody-producing plasma cells in situ, depending also on interaction with the appropriate T cells. Following antigenic stimulation lymphoid cells of the respiratory tract can produce antibody independent of the systemic antibody production.

When first formed and circulated, IgE molecules are bound to the surface of tissue mast cells and basophilic granulocytes. These cells have membrane receptors for the e-2 domain of the Fc-region of the IgE molecule (15).

Clinical Expression (Experimental)

Bronchial provocation testing (BPT) may be used to establish a definite cause-and-effect relationship if properly performed, using acceptable allergen extracts and appropriate controls. The quality of the extract used is critical. When combined with pharmacodiagnostic procedures, BPT may also provide some indication about the type of tissue responses induced by the mould extract. Such challenges may result both in immediate and in subsequent late reactions. The immediate reactions may be blocked by cromolyn and are relieved by beta-adrenergic agonists, but the late responses need corticosteroids to be abrogated.

Pepys (18), in particular, studied challenge tests in patients with allergic bronchopulmonary reactions to *Aspergillus* and *Candida*. They produced positive dual immediate and late bronchospastic asthmatic responses, often associated with systemic symptoms such as fever, malaise, leukocytosis, and eosinophilia. Some patients demonstrated late asthmatic responses without any initial immediate reaction. From the results of clinical studies alone, however, it is difficult or impossible to discern if these late inflammatory tissue reactions are the result of IgE-mediated mast cell release, IgG-immunocomplex reaction, or combinations of these and other immune and tissue mechanisms.

Responses at the Tissue Level

The immediate hypersensitivity reaction is initiated when the mast cell-bound IgE antibodies are allowed to react with the homologous allergen. To bring about a reaction, two or more IgE antibody molecules have to be bound together ("bridged") by one allergen molecule. The allergen must thus have at least two allergenic determinants accessible for the immunologic reaction. It is possible, but has not been shown, that reactions with IgG homocytotropic antibodies also depend on bridging. The immunologic bridging reaction most probably results in conformational changes in, or close to, the mast cell receptors for IgE. Cross linking of the IgE molecules by anti-IgE or of the receptors by antireceptor has about the same effect on the mast cell as linking or bridging with an allergen (15).

The mast cell serves to transform and amplify the immune reaction into one of dynamic biochemistry, which then produces various tissue effects. Mast cells are abundant in connective tissues with locations that afford membrane-bound antibodies ready access to antigens and allergens. The cells contain granules composed of histamine-heparin complexes and also contain granule-associated putative mediators that are inactive in the bound state. On activation by the allergen-IgE reaction, the granules are transported to the membrane and are discharged. The preformed mediators are released, and other bound mediators become active on solubilization.

The biochemical consequences in question are complex (9). They can, for didactive purposes, be ascribed to three stages or biochemical sequences with respect to type of reaction and time intervals (Fig. 8–4). The biochemical and biologic transitions may be diffuse and hard to distinguish clinically. The smooth muscle cells in small vessels and in the walls of the bronchi, intestine, and other organs are the effector instruments per se for the first, and also partly for the second, humoral stage of mast cell-initiated biochemical reactions. The smooth muscle cell is controlled by the autonomic nervous system with one effect (contraction of bronchioles or dilation of small blood vessels) induced by cholinergic stimulation from the parasympathetic fibers (19). In addition

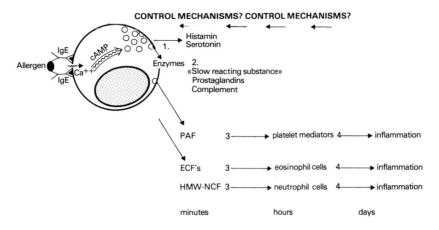

Fig. 8–4.   The mast cell serves as a transformer and amplifier for the immune reaction (bridging of IgE antibodies by the allergen) into complex biochemical sequences.

to receptors for the agents regulating muscle tone through autonomous nerve stimulation, the smooth muscle cell has receptors for other biochemical agents (Fig. 8–5). The tone of the muscle cell is a product of the sum of receptor binding and intercellular energy transmission.

The regulation of smooth muscle tone or reactivity may be impaired for different unknown reasons, as for instance in the bronchi in some patients with bronchial asthma. When the smooth muscle tissue is imbalanced in the cholinergic direction (hyperreactive), less stimulation by the mast cell-derived mediators is needed to elicit a reaction in the effector cells. Furthermore, even the presynaptic nerves of the autonomic nervous system may be influenced by biochemical substances released or activated in the environment, with modulation in one or the other direction of the regulatory nerve impulses. Agents capable of stimulating cellular adenylate cyclase decrease IgE-mediated mediator release and counteract effects of the released agents on smooth muscle.

The smooth muscle cells may also act for agents that are active during the third cellular stage of the mast cell-initiated reaction, but probably only to a rather modest degree as compared to the inflammatory changes exerted by other tissue components. Regulation of the responsiveness and responses in question is also a matter of biodegradation of the mediators being studied.

The humoral primary and secondary stages of mast cell-initiated tissue effects are followed by a rather complex set of inflammatory tissue changes. These occur when mediators from the cells are called to the site of the allergic reaction by chemotactic and activating factors, in a complex interplay with the primary and secondary stage mediators. The latter may have prepared the ground by altering local vascular tone and

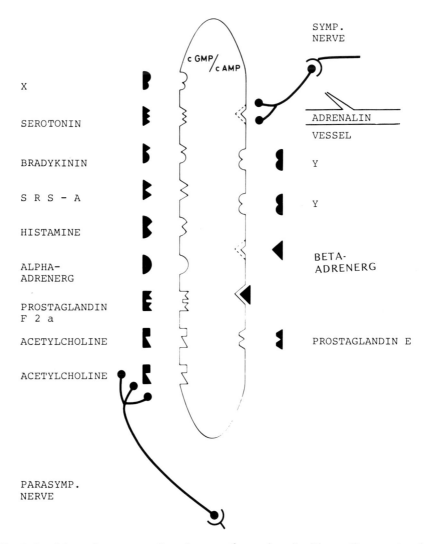

Fig. 8–5. Schematic representation of a smooth muscle cell with specific receptors for several agonists important in allergy and asthma. X and Y indicate that unidentified mediators may also be important.

permeability, allowing further extravascular concentration of serum components at the site.

The manifestations of these tissue reactions depend on the degree of immunologic sensitization and on the degree and site of allergen introduction. They depend also on the total immunogenetic potential and previous immunologic experiences of the host. Furthermore, genetic disposition for enzymes and the total biochemical balance at the moment is decisive, because these determine the availability, inductivity, and metabolism of the biochemical agent, their interactions, and binding to receptors. The same background determines the function of receptors, interactions between agonists and receptors, and effector cell responsiveness. Extremely complex interplays between enhancing and inhibitory systems and regulatory mechanisms are thus involved.

There is no evidence suggesting that immediate hypersensitivity reactions to fungal agents are different from those elicited by other inhalant allergens. Research and clinical practice concerned with fungi have, however, met more problems than encountered with most other allergen sources. These are mostly a result of the very nature of the fungi and of the fungal spores themselves. Undue consideration of this or lack of knowledge has resulted in the marketing and use of unsatisfactory preparations for research and clinical use. Our understanding of the molecular events, tissue responses, and clinical expressions of allergy to fungal agents is derived from studies employing different kind of fungal material. It seems appropriate to discuss some of the problems encountered in this respect.

FUNGAL ALLERGENS AND ALLERGEN EXTRACTS

Fungal allergen extracts are used both to diagnose allergy and to treat hyposensitization. Diagnostic tests are performed to support a diagnosis based on a case history and to provide further data necessary for initiating hyposensitization. Their use in allergic individuals is widespread, and probably will remain so in the years to come. The diagnostic use of extracts includes *in vivo* tests, such as various forms of skin testing and provocation tests, and *in vitro* tests, such as the radioallergosorbent test (RAST).

Diagnostic testing should aim at the highest possible level of precision and should not harm the subject. The quality of extracts should be set accordingly. Immunotherapy implies repeated administration of relatively high doses of extracts. Although the mechanism of hyposensitization is not fully understood, it is clear that the treatment induces some kind of alteration of the immune system. This sets even higher demands for quality of the extracts employed (2,6).

From several reasons (discussed more in detail by Salvaggio and Aukrust) (20), fungal extracts seem to be of lower and more variable quality than allergen extracts from nonfungal sources.

Variability

In regard to moulds, everything that can vary will vary. This is the single most important reason for the variable quality of commercial mould extracts. Fungi exist in a high number of species and in an even higher number of strains. It has also been suggested that fungi have a high mutation rate; they often mutate after lyophilization and when subcultured repeatedly. Different isolates of the same species will often prove to represent different strains, with more or less different properties and characteristics. In the case of *Cladosporium herbarum,* the variability among different strains is evident as shown in Table 8–1.

Growth factors, such as temperature, time, availability of oxygen, size of inoculum, and the composition of the medium, including pH, nitrogen source, and carbon source, are probably important in determining allergen content. Little information exists on the ideal culture conditions needed to obtain adequate raw material with optimal allergen content. It may also be difficult or impossible to reproduce identical batches of raw material, even when experimental conditions are kept as constant as possible.

A classic question in mould allergy is whether or not spores (conidia) or mycelia provide the best source of allergens (see Chapter 6, "Mould Allergens"). Both mycelium and spore-containing material, as well as growth medium, are possible starting sources for extraction. Growth medium may contain considerable allergenic activity but it may be difficult to prove that it is qualitatively representative of allergens in the mycelium-spore-containing mat. Because of enzymatic degradation, allergens in the medium may be partially degraded and heterogeneous, thus complicating subsequent purification.

Extracts produced by the same strains of an organism grown under identical culture conditions and for the same length of time may vary markedly in potency. It is thus recommended that standardized pools of a wide variety of strains be employed to ensure the presence of a sufficient number of allergens representing a particular species.

TABLE 8–1. Comparison of Ten Extracts of Different Isolates of *Cladosporium herbarum*

Strain No.	RAST (% Uptake of ^{125}I-Anti-IgE)	Ag-32 (% of Reference)	Ag-54 (% of Reference)	Protein (% of Dry Weight)	Carbohydrate Hexoses (% of Dry Weight)	No. of Precipitates in CIEP
Reference	39.6	100	100	14.5	26	32
1	40.0	1000	65	8.8	85	16
2	39.5	100	50	9.0	83	22
3	35.8	100	50	3.9	42	12
4	34.8	30	60	5.2	36	23
5	33.2	120	30	3.2	54	22
6	32.6	5	45	3.4	68	24
7	32.3	100	40	4.0	36	30
8	27.0	30	30	3.1	42	24
9	21.2	30	1	2.2	89	8
10	7.6			1.2	74	5

Complexity

One of the most notable properties of moulds is their complexity. Fungi are whole organisms capable of growing under different and often continuously changing nutritional and environmental conditions. The antigenic complexity of moulds is well demonstrated in crossed immunoelectrophoresis (CIEP), a two-dimensional electrophoretic technique that employs antiserum against an appropriate crude antigenic extract to obtain a high number of immunoprecipitates (4). In rabbits immunized with different *Cladosporium herbarum* extracts, approximately 60 precipitates have been obtained by the use of CIEP (Fig. 8–6). A similar complexity has been demonstrated in other fungi.

The number of distinct allergens present in crude mould material is often high, as demonstrated by crossed radioimmunoelectrophoresis (CRIE). This technique is based on IgE binding to CIEP immunoprecipitates, followed by incubation with ^{125}I-anti-IgE and autoradiography on

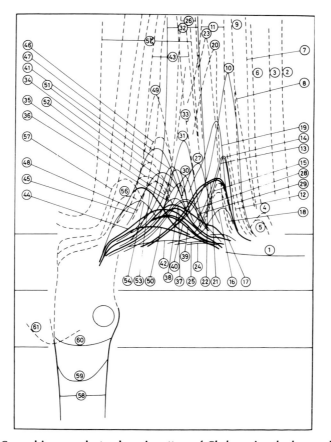

Fig. 8–6. Crossed immunoelectrophoresis pattern of *Cladosporium herbarum* showing 61 precipitates (4).

x-ray photographic film (4,20). Allergens may be defined as major, intermediate, and minor, depending on their IgE-binding in CRIE (5). *C. herbarum* seems to contain at least one major allergen, 10 intermediate allergens, and 25 minor allergens (5).

In addition, each allergen may be heterogeneous. This is illustrated by the major allergen in *C. herbarum* (Ag-32), which has been shown to exist in five molecular variants that differ only in isoelectric point but are antigenically and allergenically identical (8). Similar heterogeneity with respect to isoelectric point has been demonstrated in isolated allergens from *Alternaria* (7,21).

Cross Reactivity

Because of the great number of fungi of concern to the allergic patient, it is important to know the degree of allergenic cross reactivity among various species. In clinical practice, mould-sensitive patients often demonstrate broad patterns of positive skin reactivity to fungi of many species. This has led to the assumption that these fungi cross react. From our experiences a more likely explanation is that mould-allergic individuals often display parallel and independent multiple sensitivity to different species. This has been confirmed by others (20).

Toxicity

Fungi are known for their ability to produce potent toxins. Certain fungi have developed an ability, under special environmental conditions, to change their metabolism. Some metabolites thus produced may be toxins, which is not necessarily an advantage for the fungus. Fortunately, most mycotoxins that are potentially important are of low molecular weight (<1000 daltons), and are easily removed from allergen extracts. Some mycotoxins such as aflatoxin, ochratoxin, and patulin are well characterized, and sensitive tests exist for their detection. Such assays should be employed when dealing with fungi that are known to produce these toxins. In addition, unknown toxins may be produced. Consequently, general toxicity tests must be performed on all extracts for human use. Fungi are also potential producers of carcinogens or mutagens. For a crude extract of fungal antigens to be used in humans, it should be documented that the material is negative in a test for mutagenicity.

Potency

Biologic standardization is a means for defining the potency of allergen extracts by skin prick testing in groups of patients with a relevant sensitivity to the allergen in question (3). Using this method it is possible to compare the potency of extracts, expressed as biologic units per ml (BU/ml), prepared from different allergen sources. Crude extracts of *Cladosporium herbarum* are weaker than pollen extracts made in a similar manner. The most concentrated commercial *C. herbarum* extracts (1/20 and 1/100, weight/volume) investigated by Croner et al. and by Dreborg (12,13), had a potency that was insufficient for diagnosis of most patients. The same patients were positive when diagnosed with skin prick tests, conjunctival provocation tests and bronchial challenge tests when a highly purified, biologically standardized, and freeze-dried allergen preparation was used.

Insufficient concentration of allergens, therefore, in addition to the possible absence of some allergens in commercial extracts, may explain the high frequency of false negative skin prick tests observed (12,13). This may even explain the apparent lack of efficacy of immunotherapy with conventional fungal extracts. It may be practically impossible to reach a dose that is even close to an optimal maintenance dose. For comparison, the maintenance dose aimed at in immunotherapy with other allergen extracts is often 100- to 1000-fold higher than doses that could be reached using the highest concentrations of the commercial *C. herbarum* extract studied. At the tissue level the contents of irritants in mould extracts may induce unwanted effects. Diagnostically, these may result in false positive reactions, particularly in intradermal and bronchial provocation tests (11). Extract irritancy may also elicit local reactions during immunotherapy, giving the impression that the highest tolerated (maintenance) dose has been reached.

FUTURE RESEARCH

An important objective for future consideration in the field of fungal allergy is the selection of species used for the clinical routine or for research studies. The choice of species and strains used today seems to a large degree to be arbitrary. More data on cross reactivity and shared allergenic determinants are needed. The clinical significance of multiple mould sensitivities and their consequences for immunotherapy must be established. It may also be worthwhile to investigate the possible combined effect of toxins and allergens on sensitization.

To obtain precise and reliable information about the role and mechanism of immediate hypersensitivity to moulds on the molecular, cellular, and tissue levels, we recommend the following:

1. Adequate and representative mould preparations should be obtained.
2. All studies that have not been performed with such material should be repeated.

REFERENCES

1. Aas, K.: What makes an allergen an allergen? Allergy, **33**:3, 1980.
2. Aas, K.: The optimal allergen preparation for clinical use. *In* Clinical Immunology and

Allergology. Edited by C. Steffen and H. Ludwig. Amsterdam, Elsevier–North Holland Biomedical Press, 1981.
3. Aas, K., et al.: Standardization of allergen extracts by appropriate methods. The combined use of skin prick testing and radioallergosorbent test. Allergy 33:130, 1978.
4. Aukrust, L.: Crossed radioimmunoelectrophoretic studies of distinct allergens in two extracts of *Cladosporium herbarum*. Int. Arch. Allergy Appl. Immunol., 58:371, 1979.
5. Aukrust, L.: Allergens in *Cladosporium herbarum*. *In* Advances in Allergology. Edited by A. Oehling et al. New York, Pergamon Press, 1980.
6. Aukrust, L.: Characterization and purification of allergen extracts. *In* Diagnosis and Treatment of IgE-Mediated Diseases. Edited by S.G.O. Johansson. Uppsala, Excerpta Medica, 1981.
7. Aukrust, L., et al.: Partial purification and characterization of an allergen (Ag-8) and quantification of this allergen in eight strains of *Alternaria alternata*. Proceedings of the European Academy of Allergology and Clinical Immunology, Clermont-Ferrand, 1981.
8. Aukrust, L., and Borch, S.M.: Partial purification and characterization of two important *Cladosporium herbarum* allergens. Int. Arch. Allergy Appl. Immunol., 60:68, 1979.
9. Austen, K.F., Wasserman, S.I., and Goetzl, E.J.: Mast cell-derived mediators: Structural and functional diversity and regulation of expression. *In* Molecular and Biological Aspects of the Acute Allergic Reaction. Edited by S.G.O. Johansson, K. Strandberg, and B. Uvnaes. New York, Plenum Press, 1976.
10. Blackley, C.: Experimental Research on the Causes and Nature of *Catarrhus nestivus* (Hay Fever or Hay Asthma). London, Baillière, Tindall, & Cox, 1873.
11. Browning, W.H.: Mold fungi in the etiology of respiratory allergic disease. II. Mold extracts—a statistical study. J. Allergy, 14:231, 1943.
12. Chroner, S., et al.: Skin prick test (SPT) results in patients with possible mould hypersensitivity. (abstract). Presented at a meeting of the European Congress of Allergology, London, 1982.
13. Dreborg, S.: *In* Experience with Spectralgen/Pharmalgen: A New Kind of Allergen Preparation. Edited by B.N. Chandler. Amsterdam, Excerpta Medica, 1983.
14. Hogg, J.C., et al.: Pathologic abnormalities in asthma. *In* Asthma: Physiology, Immune Pharmacology and Treatment. Edited by L.M. Lichtenstein and K.F. Austen. New York, Academic Press, 1977.
15. Konig, W., Theobald, K., Moller, G., Pfeiffer, P., and Bohn, A.: The IgE receptor. *In* New Trends in Allergy. Edited by J. Ring & G. Burg. Berlin, Springer-Verlag, 1981.
16. Marsh, D.G.: Allergens and the genetics of allergy. *In* The Antigens. Vol. 3. Edited by M. Sela. New York, Academic Press, 1975.
17. Nordvall, S.L., et al.: Characterization of the mouse and rat IgE antibody responses to timothy pollen by means of crossed radioimmunoelectrophoresis. Allergy, 37:259, 1982.
18. Pepys, J.: Hypersensitivity diseases of the lungs due to fungi and organic dusts. *In* Monographs in Allergy. Vol. 4. Edited by S. Karger, New York, 1969.
19. Richardson, J.B.: The neural control of human tracheobronchial smooth muscle. *In* Asthma: Physiology, Immunopharmacology and Treatment. Edited by L.M. Lichenstein and K.F. Austen. New York, Academic Press, 1977.
20. Salvaggio, J., and Aukrust, L.: Mold-induced asthma. J. Allergy Clin. Immunol., 68:327, 1981.
21. Yunginger, J.W., Jones, R.T., Nesheim, M.E., and Geller, M.: Studies on *Alternaria* allergens. III. Isolation of a major allergenic fraction (ALT-1). J. Allergy Clin. Immunol., 66:138, 1980.

Chapter 9

INCIDENCE AND CLINICAL CHARACTERISTICS OF MOULD ALLERGY

William A. Howard

Airborne mould spores are widespread in nature, so it is not surprising that spores should be scrutinized as possible causes of disease in humans. Allergy to moulds has been presumed, based largely on clinical observations and on the presence of positive skin test reactions to certain mould antigens. There is little evidence as yet, however, to demonstrate the cause-and-effect relationship of moulds to the many forms of clinical allergy. The difficulties encountered have been enumerated by Salvaggio and Aukrust (18). These include the following:

1. Selection of fungus antigens has varied widely among different investigators.
2. Preparation and standardization of mould extracts have also varied widely, and quality and potency have not been adequately monitored.
3. Only a few species of mould spores have been studied in detail.
4. Identification and classification of fungi have been difficult.
5. Fungal extracts have been prepared in the same manner as for pollens in spite of the fact that mould cell walls contain only chitin and cellulose, while pollen cell walls contain much extractable protein.
6. There has been a lack of characterization of fungal antigens as compared to what has been accomplished for some pollens.

7. There is no single well-defined mould season.
8. The extent of uniqueness and cross reactivity among fungi remains to be determined.

With these pertinent facts in mind, it is not surprising that a high correlation between mould sensitivity and clinical symptomatology has not been obtained. Regardless of these obstacles, however, it appears reasonable to assume that allergic responses to moulds do occur, and that they may mimic the broad spectrum of allergic manifestations associated with other common inhalant allergens.

HISTORICAL BACKGROUND

Moulds have been recognized for centuries, although identification and classification have been a relatively recent accomplishment. They are ubiquitous in nature, but their relationship to human ailments has evolved only slowly. Involvement of moulds in asthma may have been suggested by Moses Maimonides in the twelfth century when he described the frequent occurrence of wheezing in damp weather. In 1726 Sir John Floyer (8) noted the development of violent asthma in a patient who had just visited a wine cellar (8). Blackley, in 1873, suggested the association of species of *Chaetomium* and *Penicillium* with attacks of bronchial catarrh. Van Leeuwen (21), in 1924, again noted the relationship of climate to asthma, and also pointed out a definite correlation between mould spores and asthma. In the same year the first case of asthma due to a fungus (wheat rust) was reported.

Other investigators began to report cases of asthma caused by moulds, and demonstrated positive cutaneous reactions and positive passive transfer tests to mould extracts in these individuals (2,4,6). Challenge with nasal insufflation of mould spores was found to precipitate an attack of asthma in the sensitive individual.

Feinberg and Little (7) became interested in the role of moulds in clinical allergy, based in part on the developing literature but more specifically on the observation that there were large numbers of mould spores on slides used for pollen identification and counting. They also noted that hay fever symptoms often appeared at times when pollen counts were low but when mould spores were present in large numbers. They suggested that basidiospores might be a significant factor in respiratory allergy, an observation that has recently been the subject of intensive investigation.

From these beginnings information and experience have increased, and there is now a substantial body of evidence linking moulds to many manifestations of clinical allergy.

INCIDENCE

Moulds have a worldwide distribution and, with the many thousands of species recognized, the magnitude of the task of defining those that may be significant in human disease can be readily appreciated. As discussed in other chapters, much of our knowledge concerning moulds has come from air sampling, identifying those mould spores that occur most frequently in our environment. It has been logically assumed that those species found in the greatest number are most likely to be identified with disease states in humans. Thus, preparation and use of mould extracts have been limited to those species most abundant in the environment and most easily cultivated. Matching these species to human disease has occurred chiefly through skin testing of individuals known to have had the opportunity for exposure to the moulds being tested.

In this manner certain mould genera have been recognized as those most often producing hypersensitivity skin test reactions in allergic individuals. These include *Alternaria, Cladosporium, Helminthosporium, Fusarium, Penicillium, Phoma, Aspergillus, Rhizopus,* and *Mucor.* Other members of *Aureobasidium,* fungal classes such as the Basidiomycetes, are being investigated as possible causes of fungal allergy in humans (12). Mould spore counts by use of the Rotoslide method along the eastern seaboard in the vicinity of Washington and Baltimore have shown extremely high numbers of *Alternaria* and *Cladosporium.* Yearly totals may reach into the hundreds of thousands, greatly exceeding comparable ragweed counts. Gravity sampler counts in the Washington DC, area indicate that spores of *Alternaria* and *Cladosporium,* as well as others, can be found in relative abundance throughout the year, although there may be seasonal increases in the spring and during the months of September and October (1).

It is difficult to obtain information on the incidence of specific mould sensitivity in the allergic population. Mazar et al. (13) studied 97 asthmatics and found that 38 had antibodies to one or more inhalant fungal allergens as measured by immunodiffusion or by crossed immuno-electrophoresis, while only 2 of 39 patients with chronic obstructive pulmonary disease had similar findings. Significant positive skin test reactions to one or more inhalant allergens were found in 66 of the 97 asthmatics. Of this group, 15 reacted to *Cladosporium,* 12 to *Alternaria,* and 8 to *Aspergillus.* The question of clinical correlation was not addressed. Of 110 randomly selected pediatric patients in the Washington, DC, area with symptoms of rhinitis and/or asthma, 92 (83%) has significant (greater than 5 mm) reactions to one or more inhalant allergens (19). Most of these (80 of 92) reacted to house dust, an expected finding in childhood allergies. Pollen reactions were noted in 48 of the 92 (52%), while mould reactions accounted for 52 of the 92 (56%). Of the moulds tested, *Alternaria* reacted most commonly with 38 of the 52 (73%), followed by *Helminthosporium* with 20 of 52 (38%), and *Cladosporium* with 14 of 52 (27%).

In spite of the many gaps in our knowledge of moulds and their relationship to clinical allergy, there seems no doubt that at least some moulds are related to respiratory allergy in a significant proportion of atopic individuals

So far we have considered only outdoor airborne moulds. Moulds are also found about the house in varying concentrations, depending on environmental temperatures and degree of humidity. Laundry rooms, basement storage areas, crawl spaces, bathrooms, shower stalls, refrigerators, and stored foods (onions, potatoes) also promote mould growth. When present, this contributes to increased spore concentration inside the house. Mould spores originating in damp basements may be carried throughout the house by convection air currents, and may appear in sufficient concentration in other parts of the house to be a factor in the production of symptoms. Mould plates exposed in strategic areas indoors may be necessary to correlate specific exposure to positive skin test reactions and clinical symptoms. On the other hand, there is no hard evidence that ingestion of mould spores is a significant factor in the production of symptoms in mould-allergic patients.

MECHANISMS OF ALLERGIC RESPONSES

Typical allergic rhinitis, conjunctivitis, and asthma clearly have been produced by exposure to fungus spores, but no such association has been demonstrated in urticaria, angioedema, or atopic dermatitis. Positive skin tests to moulds have been demonstrated regularly in atopic patients, and several investigators have correlated these findings with positive bronchial inhalation challenges (3,20). Bruce (5) has suggested that the carefully performed skin test may be as diagnostically useful as the bronchial challenge in confirming the etiology of seasonal asthma. These findings are entirely compatible with a type I (Coombs and Gell) IgE-mediated immune response. Individuals with positive skin tests and positive inhalation challenges have been shown to have lower threshold responses to both methacholine and histamine. Those with negative inhalation challenge, irrespective of skin test reactivity, did not show these low thresholds. These data are consistent with the concept that cholinergic factors contribute to a positive inhalation challenge response (20).

Many patients with allergic asthma have a biphasic response to inhalation challenge with appropriate antigens (16). The immediate reaction occurs within 10 minutes, with a significant fall in the first second expiratory volume (FEV_1) that returns to prechallenge levels within 1 to 3 hours. This may be followed by more severe airway obstruction 6 to 12 hours after challenge. Early onset challenge-induced asthma subsides rapidly, either spontaneously or with minimal treatment. The late response is more resistant to treatment, lasts longer, and is associated with a greater degree of lung hyperinflation. This response has been noted with many different allergens, suggesting that this pattern may be relevant to the natural behavior of asthma, and that it may be more important than the immediate response (17).

Early responses are associated with the appearance of histamine and neutrophil chemotactic activity (NCA), all inhibited by the prior administration of cromolyn and suggesting involvement of mast cells. The pathogenesis of the late reaction is unknown, however, although late mast cell activity has been suggested. Nagy et al. (15) measured NCA and found that activity was increased after both early and late reactions. If only an early response occurred, there was no late rise in NCA. NCA did not appear after bronchial challenge with histamine and methacholine, suggesting that NCA was not a function or consequence of bronchospasm alone. This led the investigators to believe that mast cell degranulation is associated with both early and late responses, although mediators other than histamine and NCA may be involved. Late reactions occurred in the absence of precipitins in the sera. These findings do not necessarily eliminate the possibility of a type III reaction in the late responders, because immune complex deposit also may have occurred in these subjects.

These reactions are not to be confused with the delayed development of pneumonitis with dyspnea noted in hypersensitivity pneumonitis. When asthma is not present there is no evidence of a type 1 reaction, and clinical manifestations only develop 4 to 8 hours after exposure to the causative antigen. The presence of precipitins in the sera of some mould-sensitive patients has been noted, but these do not appear to alter or add to the tissue changes of immediate hypersensitivity (19). In some instances, such as in bronchopulmonary aspergillosis, a combined type 1 and type III reaction occurs with dual skin reactions and with asthma associated with pulmonary infiltrates, mucoid impaction, and central bronchiectasis.

Certain yeasts, such as *Candida*, regularly produce a type IV allergic response following skin or gastrointestinal involvement, and a type I reaction need not precede such a response. The classic delayed response to *Candida* is used as a test for the presence of a normal cell-mediated immune response.

Although there is some evidence of cross reactivity among many genera of moulds, these must be considered as preliminary studies and, for practical purposes, the clinician will be better served by using the best individual extracts that can be obtained.

CLINICAL MANIFESTATIONS

Moulds will grow almost anywhere there is sufficient moisture, and many species will grow in a fairly wide temperature range, although some species may require fairly specific temperature ranges to survive. Moulds are most numerous from early spring throughout the fall months, diminishing in numbers with falling temperatures. Such changes will have minimal effect on mould growth within the home, in which humidity may be the principal stimulant to mould proliferation. The heaviest appearance of mould spores tends to coincide with the pollen seasons, which are often associated with damp and changeable weather. It is not surprising therefore, for mould and pollen allergy to be confused, or to occur together in the same patient.

> A. J., 11 years of age, spent the summer in a boy's camp on a lake in New Hampshire, where he noted no allergy problems. On return to the Washington, DC, area on Labor Day, he developed acute nasal and eye symptoms and findings characteristic of hay fever. Ragweed was assumed to be the cause, but skin and conjunctival tests to ragweed were negative. A strongly positive wheal-and-flare reaction with pseudopods was obtained with *Alternaria* at 100 PNU/ml, at a time when *Alternaria* spores were plentiful on exposed slides, allowing a presumptive diagnosis of mould hay fever.

In the absence of a seasonal pattern for moulds, the environmental history becomes of paramount importance in establishing a diagnosis of mould allergy, as well as in establishing a correlation between IgE-mediated skin test results and clinical symptoms, and prevalence of moulds in the environment. Clinical characteristics of mould allergy in humans vary widely, ranging from the usual manifestations of atopic disease to those of a more invasive nature, such as bronchopulmonary aspergillosis, hypersensitivity pneumonitis, and mucocutaneous candidosis. The latter problems are covered in other portions of the text (see Chaps. 7, 12, and 13); this discussion is concerned primarily with atopic disease.

ALLERGIC RHINITIS

The classic forms of nasal allergy, hay fever, intermittent allergic rhinitis, and perennial allergic rhinitis, may occur in the susceptible individual after exposure to appropriate mould spores. The symptoms and signs may be indistinguishable from those due to dusts, pollens, or danders. Persistent or intermittent nasal obstruction, mouth breathing, nasal discharge, nose rubbing, and postnasal drip are characteristic of perennial involvement due to moulds, and greatly resemble the persistent and prolonged findings in nasal allergy caused by dust. Bouts of sneezing and eye irritation with lacrimation, redness, and itching may appear in more acute episodes. A hawking, irritating cough may be present, especially at night, with or without a postnasal drip. Nasal symptoms are likely to be of lesser intensity than in pollen hay fever, persist for more prolonged periods of time, and are more or less intermittent in nature. The occasional patient may show acute exacerbations after heavy exposure, especially after raking or playing in damp or decaying leaves, or after returning to a heavily mouldy area after being in a relatively mould-free environment.

In patients with rhinitis look for associated infection, especially sinusitis, which may be superimposed on the allergic base. A deviated septum may mimic nasal allergy, or may make an existing problem worse, as will any injury to the nose. The presence of enlarged adenoids may contribute to nasal obstruction and difficulty in breathing through the nose when nasal passages appear clear and open. Cystic fibrosis may present with marked nasal obstruction and polyposis, as may aspirin intolerance. Cultures of purulent nasal secretions will shed some light on the role of infection, and on a rare occasion a fungus such as *Candida* may be isolated in the absence of tissue invasion.

> J.P., 2 years of age, presented with a heavy purulent nasal discharge, unresponsive to local or systemic antibacterial therapy. Bacteriologic cultures showed no significant pathogens, but cultures for yeasts produced a profuse growth of *Candida albicans*. Therapy eventually resulted in improvement of the local process, but the child later developed respiratory allergies, with severe asthma, due to both dust and mould.

Bronchitis

Bronchial irritation with cough, in the absence of clinical asthma, may occur as it does with other inhalant allergies. There may or may not be an associated nasal problem, and a true allergic bronchitis is presumed to exist. Most of these patients will demonstrate some degree of reversible airway obstruction, with decreases in peak flow rates and in FEV_1, suggesting a latent or silent asthma. These same individuals may have exercise-induced bronchospasm with appropriate clinical findings of shortness of breath or wheezing and a fall in FEV_1, whether or not there is any change in the forced vital capacity (*FVC*).

Asthma

Classic allergic asthma also occurs in the mould-sensitive patient, and, although not usual, has been observed in the absence of dust or pollen problems. In children, moulds may be a more significant cause of asthma than pollens in some geographic areas, while the reverse appears to be true in adults, Wheezing may accompany a flare of nasal allergy due to moulds, or may be a primary response. Attacks are ordinarily introduced by nasal symptoms that suggest a cold, rapidly followed by a cough of gradually increasing severity and frequency. Wheezing may follow within hours or a day or two. Fever is not present unless there is an associated respiratory infection. The delayed or second flare of asthma noted to occur with some bronchial challenges is difficult to evaluate in the clinical situation, especially in mould-sensitive patients in whom asthma develops more slowly and is more persistent. This is not to be confused with the delayed development of pneumonitis with dyspnea noted in hypersensitivity pneumonitis, in which asthma is not present and clinical manifestations, including fever and malaise, only develop 4 to 5 hours after exposure to the causative agent.

Respiratory symptoms due to moulds may be differentiated from those resulting from other inhalant allergens by history and by skin test response or some other measurement of specific antibody. The radioallergoabsorbent test (RAST) may be useful in some situations in which direct skin testing is not feasible, but usually does not exceed the direct skin test in sensitivity. Both techniques suffer from the same problem, lack of a sufficiently refined and antigenically defined extract that will be truly diagnostic.

Mould reactions may coexist with sensitivity to pollens, dusts, and danders. In addition, especially in children, foods may play a role in allergic respiratory involvement.

Asthma and allergic bronchitis in the mould-sensitive individual must be differentiated from other possible causes of episodic or persistent wheezing and changes in pulmonary function. Conditions to be considered vary with the age of the patient. In the very young, laryngotracheomalacia may mimic the wheeze of asthma. Chronic or recurrent viral respiratory infections may be accompanied by wheezing, with or without an associated allergy. Cystic fibrosis may occur with wheezing and recurrent respiratory infections. An increased incidence of allergy to *Aspergillus* has been claimed but not confirmed in cystic fibrosis patients, although the mould may be cultured from pulmonary secretions. A foreign body must always be kept in mind in any child with cough and wheeze, especially if of sudden onset. Croup and epiglottitis may cause noisy or difficult breathing, usually inspiratory in character. In some children recurrent episodes of spasmodic croup may be followed by wheezy breathing, and may be associated with the later development of allergic asthma.

In older individuals chronic bronchitis and chronic obstructive pulmonary disease must be considered in the differential diagnosis of asthma. Other considerations include any pulmonary lesion that might produce endobronchial obstruction or extrabronchial compression, including tuberculosis, sarcoid tumor, or fungal disease. Parasitic diseases with a pulmonary phase, such as ascariasis and filariasis, may also induce wheezing in association with a pronounced eosinophilia. At times it may be difficult to differentiate allergic asthma from asthma of cardiac origin or from nonallergic asthma. Pulmonary emboli accompanying cardiac disease must also be considered.

COMPLICATIONS

The complications of mould allergy differ little from those of allergies due to other agents. Evidence of invasive disease with moulds, both locally and systemically, may be seen, especially when there is evidence of some disturbance in the immune mechanism. Bronchopulmonary aspergillosis is the most frequently recognized offender, with dual skin test reactions and destructive pulmonary involvement. Fungus balls have been reported to occur with fungi such as *Candida,* but these are rare and are easily recognized by roentgenography.

A large pulmonary mycetoma with cerebral metastases due to *Curvularia* has been reported in a child (11). A positive direct skin test was present but cell-mediated response to the organism was lacking although there was a normal delayed-type response to both *Candida* and *Histoplasma,* suggesting that, although the child had an apparently normal cell-mediated response to fungal antigens, there was a specific defect with respect to *Curvularia.* Other types of invasive disease due to mould have been recognized, including endocarditis (14), ocular infections (9), and destructive sinus infections. These are rare, and do not appear to be related to immunologically mediated responses to moulds.

In summary, moulds are ubiquitous in the environment and, although proof is not always present, there is no doubt that many of them are causally related to human allergic disease. The spectrum of immediate allergic responses does not differ significantly from that of reactions to other allergens. Mould allergy is frequently associated with delayed-type hypersensitivity. These latter characteristics give their one distinctive clinical feature to moulds: a tendency to be invasive and to produce a variable spectrum of systemic disease. This suggests a more careful approach to the diagnosis of mould hypersensitivity, and will at times make treatment somewhat more complicated. For the long term, patients with immediate IgE-mediated mould symptoms do not appear to be at any greater risk of complications than those with allergic responses to other common allergens.

REFERENCES

1. Al-Doory, Y., et al.: Airborne fungi and pollens of the Washington, D.C., metropolitan area. Ann. Allergy, **43**:360, 1980.
2. Bernton, H.S., and Thom, C.: The importance of moulds as allergic excitants in some cases of vasomotor rhinitis. J. Allergy, **4**:114, 1924.
3. Bronsky, E.A., and Ellis, E.F.: Inhalation bronchial challenge testing in asthmatic children. Pediatr. Clin. N. Am., **16**:85, 1969.
4. Brown, G.T.: Hypersensitivity to fungi. J. Allergy, **7**:455, 1936.
5. Bruce, C.A.: Quantitative bronchial inhalation challenge in hay fever patients. J. Allergy Clin. Immunol., **56**:331, 1975.
6. Cadham, F.T.: Asthma due to grain dusts. JAMA, **83**:27, 1924.
7. Feinberg, S.M., and Little, H.T.: Mould allergy. Its importance in asthma and hay fever. Wis. Med. J., **34**:254, 1935.
8. Floyer, Sir John: Violent Asthma after Visiting a Wine Cellar. A Treatise on Asthma. 3rd. Ed. London, 1726.
9. Gugnaria, H.C., Gupta, S., and Talwan, R.S.: Role of opportunistic fungi in ocular infections in Nigeria. Mycopathol., **65**:155, 1978.
10. Howard, W.A.: Personal observations, 1982.

11. Lampert, R.P.: Pulmonary and cerebral mycetomas caused by *Curvularia pallescens*. J. Pediatr., **91**:603, 1977.
12. Lopez, M., Salvaggio, J., and Butcher, B.: Allergenicity and immunogenicity of Basidiomycetes. J. Allergy Clin. Immunol., **57**:480, 1976.
13. Mazar, A., et al.: Antibodies to inhalant fungal antigens in patients with asthma in Israel. Ann. Allergy, **47**:361, 1981.
14. Mendelsohn, G., and Hutchins, G.M.: Infectious endocarditis during the first decade of life. Am. J. Dis. Child., **133**:619, 1979.
15. Nagy, L., Lee, T.H., and Kay, A.B.: Neutrophil chemotactic activity in antigen-induced late asthmatic reactions. N. Engl. J. Med., **306**:497, 1982.
16. Pepys, J., et al.: Inhibitory effects of disodium cromoglycate in allergen inhalation tests. Lancet, **2**:134, 1968.
17. Pepys, J., and Hutchcroft, B.J.: Bronchial provocation tests in etiologic diagnosis and analysis of asthma. Am. Rev. Resp. Dis., **112**:829, 1975.
18. Salvaggio, J., and Aukrust, L.: Mould-induced asthma. J. Allergy Clin. Immunol., **68**:327, 1981.
19. Solomon, W.R., and Mathews, P.P.: Aerobiology and inhalant allergens. *In* Allergy, Principles and Practice. Edited by E. Middleton Jr., et al. St. Louis, C.V. Mosby, 1978.
20. Spector, S., and Farr, R.: Bronchial inhalation challenge with antigens. J. Allergy Clin. Immunol., **64**:580, 1979.
21. Van Leeuwen, W.S.: Bronchial asthma in relation to climate. Proc. Soc. Med., **17**:19, 1924.

Chapter 10

CRITICAL REVIEW OF DIAGNOSTIC PROCEDURE FOR MOULD ALLERGY

Peter P. Kozak, Jr.
Donald R. Hoffman

Reaginic sensitivity to mould allergens is one of the least understood of all inhalant allergies. This is due, in part, to a lack of basic knowledge of mycology by both the research academician and the practicing allergist. A critical evaluation of the medical literature dealing with mould allergy indicates only a few carefully performed studies. At present, there is a need to reevaluate prior research and to define more clearly patient environmental exposure to moulds, antigen preparation, standardization of extracts, and testing procedures.

Before evaluating the various *in vivo* and *in vitro* procedures used to diagnose mould allergy, we should review methods presently employed for production of commercial mould extracts. Stock cultures are maintained by manufacturers. Inoculates are taken from the stock culture and grown either on agar or in broth. At the end of an incubation period of up to several weeks, the mature growth (mycelial mat) is separated from the medium, dried, and ground to a powder. The extract is later prepared from this dried material, usually after some defatting. In an attempt to improve the potency and specificity of mould extracts, the Association for Mycological Investigation developed the MMP process. Instead of using the mycelial mat, the broth in which the mould was grown is concentrated, dialyzed, and then freeze-dried. This material is said to contain the nondialyzable mould metabolites, including some of the allergens. Extracts of this material are prepared in a manner similar to that for preparation of other extracts. The final extract is generally maintained in 50% glycerin or is lyophilized to preserve potency.

The lack of uniform potencies for extracts of *Alternaria* has been reported by Yunginger et al. (58). Twelve commercial extracts of *Alternaria* were evaluated by comparing protein nitrogen content (expressed as protein nitrogen units, PNU) and potencies by end-point titration, the direct radioallergosorbent test (RAST),and RAST inhibition. There was no relationship between the manufacturer's designation of concentration (PNU or weight/volume, w/v) and the actual allergenic potency as determined by skin testing or by RAST evaluation, even from the same manufacturer. End-point titrations showed up to a 3000-fold difference between the extracts, with the RAST inhibition differing by more than 200-fold. Two different patterns for the RAST inhibition were noted, with significant variations within each group. These findings suggest significant qualitative differences between supposedly similar products. The authors concluded that commercial *Alternaria* extracts showed more heterogeneity than commercial extracts of either short ragweed or June grass.

Aukrust and Aas (8) have studied extracts of *Cladosporium herbarum* from two different manufacturers and compared the Phadebas RAST with their own RAST for each extract. The study involved 35 children, 15 of whom had received previous immunotherapy to *C. herbarum*. The maximum uptake by the Phadebas disc was 26.8% compared to more than 50% by both specially prepared RAST discs. Reactions of class II RAST and above were noted in 20% of patients using the Phadebas RAST disc, and 29% and 35% when RAST was prepared from Allergologisk and Allergon extracts, respectively.

Attempts have been made to improve the correlation of RAST with other *in vitro* and *in vivo* tests by increasing the amount of allergen fixed to the solid phase support (46) by either concentration (23) or partial purification (3) of the crude extract. Our group has taken a somewhat different approach. We have concentrated on obtaining spore-rich material from which to prepare extracts and for RAST studies. Hoffman et al. (26) reported results of prick skin tests and RAST comparing special spore-rich, mycelial-rich, and spore-specific extracts with Hollister Steir *Alternaria* MMP. The only purification needed to prepare the RAST reagents was diafiltration to remove low molecular weight materials. In group III, composed of 44 patients, the RAST and skin tests were positive and concordant in 39 of 39 patients tested with mycelial-rich extracts, 44 of 44 tested to spore-rich (crude spore) extracts, and 43 of 44 tested to spore-specific *Alternaria* extracts. Forty-one patients gave positive prick tests and RAST to a related genus, *Ulocladium*. Three patients had positive RAST to mycelial-rich extract while the skin tests were negative. One patient had a positive skin test to the spore-specific extract but a negative RAST. Approximately half the patients studied showed a stronger reaction to the *Alternaria* spore-specific extract. Several commercial *Alternaria* extracts were studied and showed low amounts, if any, of the spore-specific antigens.

Aas et al. (3) compared skin test reactivity and direct RAST for three different commercial extracts of each of the following: *Alternaria, Aspergillus, Cladosporium, Mucor,* and *Penicillium.* Positive prick skin test reactions for extracts of *Cladosporium herbarum* varied from 12.5 to 86% in the same population. The authors concluded that the extracts differed not only in respect to total allergenic activity but also in concentration of individual allergens.

EVALUATION OF DIAGNOSTIC MATERIALS

The allergen extract must be appropriate and potent regardless of which *in vivo* or *in vitro* techniques are used to confirm reaginic sensitivity. Commercial mould extracts presently available for diagnosis and treatment lack adequate sensitivity and specificity. Four factors that contribute to the present variability of commercial mould extracts include the following:

1. Variability of stock cultures used to prepare commercial extracts, and their accurate identification.
2. Current use of mycelial-rich material to prepare the extracts.
3. Conditions under which moulds are grown and extracts are prepared.
4. Stability of the extracts.

Mould Identification

Many commercial mould extracts are prepared from stock cultures obtained from the American Type Culture Collection. Some manufacturers use local isolates of moulds to prepare their extracts. The Bureau of Biologics of the Food and Drug Administration leaves it up to the individual manufacturer to confirm identification of mould, regardless of the source. Yunginger et al. (58) contacted ten manufacturers and confirmed that seven extracts were prepared from strains of *Alternaria tenuis,* one from *A. alternata* (present designation for *A. tenuis*), one from *A. brassicicola,* and one from an unspeciated strain of *Alternaria.*

The problem is further compounded by mislabeling of extracts. One widely used extract of *Helminthosporium* contains *H. maydis* and *H. interseminatum.* Using the currently recommended taxonomy, the *Helminthosporium maydis* should be labeled as *Dreschslera* and the *H. interseminatum* as *Curvularia* (42). The problem is not necessarily one of deliberate misrepresentation, but reflects the changing taxonomy related to mould identification. Rather than "confuse the practicing allergist," manufacturers have continued using outdated nomenclature. Although some discrepancies continue to exist in fungal taxonomy, there is sufficient agreement to identify most important moulds accurately. Confirmation of identification by an independent authority is generally possible and is highly recommended.

Mycelial vs. Spore Extracts

Aerosampling indicates that the spore is the major mould component found in the air. Mycelial fragments are also found, but it is impossible to identify the mould genera that they represent. Most commercial extracts contain mainly mycelial components, with minimal spore content. Examination of the crude material used to prepare the extracts amply illustrates the point (Fig. 10–1). Crude material from which three different manufacturers prepare *Alternaria* extracts was compared to a spore-rich material used by Hoffman et al. (26). Only occasional spores can be identified in any of these crude commercial materials.

We have recently used the API-zym system for identification and quantitation of enzymes present in the various *Alternaria* extracts. Mycelial-rich (MREx) and spore-rich extracts (SpREx) were compared to the broth used to grow *Alternaria* (Table 10–1). Broth in which we grew the *Alternaria* mycelium contained significant quantities of acid phosphatase and N-acetyl-β-glucosaminidase, with only trace quantities of four other enzymes. The mycelial-rich extract contained additional enzymes, including alkaline phosphatase and β-glucosidase. Spore-rich extracts contained significant amounts (in excess of $1+$) of 12 of the 19 enzymes studied. All three extracts were tested 2 years after preparation. They had been stored at 4° C and gave almost identical reactions to later extracts.

The enzymes that appear to be associated with *Alternaria* mycelial extract include alkaline phosphatase, acid phosphatase, β-glucosidase and N-acetyl-β-glucosaminidase. Enzymes most closely associated with spore extracts include leucine aminopeptidase, α-galactosidase, β-galactosidase, and β-glucuronidase. Only one commercial *Alternaria* extract has an enzyme pattern similar to the spore-rich extract (extract F).

Culture and Extraction Conditions

Growth of mould cultures are affected by various conditions, including temperature, humidity, light-dark cycle, and media on which they are grown. These conditions affect sporulation and the antigen content of the material. Optimal methods for the preparation of mould extracts are not always established. Traditionally, commercial mould extracts have been prepared in a manner similar to that employed for preparation of pollen extracts. Mould spores are significantly more complex than other inhalants and may require a different extraction method. The cell wall of mould spores contains chitin, cellulose, polysaccharides, lipids, and specific proteins. Using cross immunoelectrophoresis (CIEP) Lowenstein

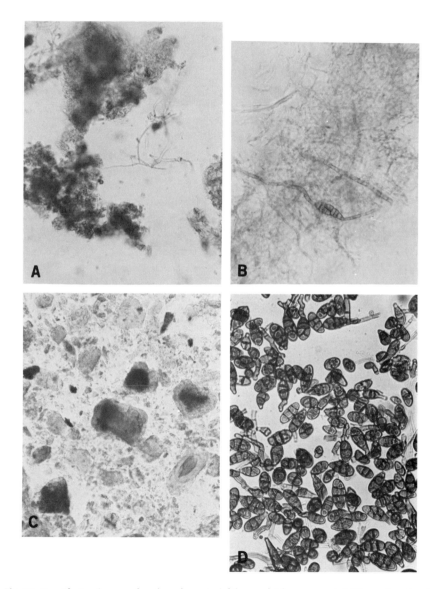

Fig. 10–1.   Photomicrographs of crude material from which commercial *Alternaria* extracts are prepared. *A*. Powdered *Alternaria*. *B*. *Alternaria* freeze-dried mycelia and spores. *C*. Powdered *Alternaria*, defatted. *D*. *Alternaria* spore-rich material.

TABLE 10–1. API-zym Assays of Different *Alternaria* Extracts

Enzyme Assayed	Special *Alternaria* Extracts			Commercial Extracts					
	Broth	Mycelial-Rich Extracts	Spore-Rich Extracts	A	B	C	D	E	F
Alkaline phosphatase	Trace	3	4			2	1	3	4
Esterase (C4)	Trace	1	3		4	2	2	2	4
Esterase lipase (C8)	Trace	1	4		3	3	4	2	4
Lipase (C14)									
Leucine aminopeptidase		Trace	5			3	2		4
Valine aminopeptidase			Trace						
Cystine aminopeptidase									
Trypsin			1						
Chymotrypsin									
Acid phosphatase	3	3	4	1	3	1	4	4	4
Phosphoamidase	Trace	1	3	1				3	2
α-galactosidase		Trace	5					1	4
β-galactosidase			5						4
β-glucuronidase		Trace	5						3
α-glucosidase		1	4		2	3		4	3
β-glucosidase		5	5		2	3	2	2	5
N-acetyl-β-glucosaminidase	3	4	5	2	2	2	2	2	5
α-mannosidase			1						
α-fucosidase									

(34) has shown between 40 to 50 arcs to *Alternaria* extracts and 57 arcs to *Cladosporium herbarum* (33,34). Although isolation and characterization of these antigens will take years to accomplish, these will ultimately lead to more potent and uniform extracts.

Stability of Extract

Deterioration of extracts probably begins shortly after preparation. Stability of extracts is in part related to storage temperature and to the presence of glycerin (9). Lyophilization also may extend extract stability but there is insufficient data at present to recommend its use.

Work is presently under way in many countries to standardize extracts and to monitor their proper manufacture and distribution. Use of high-quality extract for diagnosis will make a major contribution to our future understanding and treatment of mould allergy.

The presence of enzymes in mould extracts has not been appreciated by many. Kimura et al. (29) were able to identify carboxylic ester hydrolase in an extract of *Fusarium solani* and acetylcholinesterase, glucose-6-phosphatase, glucose-6-phosphate dehydrogenase, carboxylic ester hydrolase, and lactic dehydrogenase in extracts of *Puccinia coronatoa* (oats rust). In addition to the enzymes already identified in *Alternaria* extracts (Table 10–1), our laboratory studies have shown significant enzymes in several other spore-rich extracts and their commercial counterparts. Using the API-zym system we have been able to evaluate these extracts for only 19 enzymes. Undoubtedly other important enzymes are also present. The manner in which these proteinases and glycosidases contribute to deterioration of mould extracts has not been determined.

DIAGNOSTIC PROCEDURES

Although scattered reports of inhalation challenge studies appeared in the medical literature prior to the development of skin testing, the latter testing technique has received far greater acceptance. Various *in vivo* and *in vitro* methods for diagnosis of atopic respiratory disease have subsequently been developed, as shown in Table 10–2.

There is an abundance of literature evaluating these testing methods for many allergens, but there is relatively little information about mould antigen. We will review the more important studies in each of these areas and, where possible, highlight the data concerning mould sensitivity. Each testing method will be reviewed with emphasis on basic mechanisms, problems, advantages, disadvantages, and reliability. An overview of research into mould allergy will follow in the summation.

In Vivo Testing Procedures

Skin Testing

The skin test is a deliberate and controlled exposure to a suspected allergen conducted mainly to confirm clinical atopic sensitivity. It involves an interaction between the allergen and IgE fixed to mast cells in the skin with liberation of chemical mediators, resulting in local erythema and wheal formation. A basic premise of skin testing is that reaginic antibody will fix to the skin mast cell in a manner similar to that occurring in the various target organs. Widespread use of the skin test as a diagnostic aid began in the early part of this century. It was hoped that skin testing would separate the atopic population sensitive to an allergen from the nonallergic group. Skin testing was quickly adopted and has become the primary method for detecting etiologically important factors in atopic disease.

Discordance between clinical history and skin testing has been ap-

TABLE 10–2. Methods for Diagnosis of Atopic Respiratory Disease

In Vivo Procedures	In Vitro Procedures
Skin testing	Radioallergoabsorbent test (RAST)
Scratch and prick tests	Enzyme-linked immunosorbent
Intradermal skin test	assay (ELISA)
End point titration	Histamine release from basophils
Challenges (provocative studies)	
Nasal challenge	
Bronchial challenge	

parent for some time. This has led to the use of inhalation challenges in addition to the development of various *in vitro* tests to confirm the importance of skin test reactivity. Inhalation challenge to confirm skin test reactivity is most commonly performed in Scandinavia; immuno- therapy is frequently withheld unless there is agreement between the results of these studies.

Regardless of which skin testing procedure is employed, several prob- lems may be encountered. Variability of response is due in part to the age of the patient. Infants and young children have only a minimal whealing response to testing, and diminished responses are also noted in the elderly. Skin test responses can be blocked or reduced by the use of medications, especially antihistamines. Dermatographism is occa- sionally encountered. Suitable positive (histamine) and negative controls must be used.

Extracts used for testing should be standardized to ensure equal po- tency from different manufacturers and also to minimize batch-to-batch differences from the same manufacturer. Once extracted, the antigen should be properly stored and transported to minimize deterioration of the material. Nonspecific irritant reactions should be excluded by testing in nonatopic controls.

Scratch and Prick Test. The scratch test is performed by making a controlled superficial cut into the epidermis without drawing blood. To control the depth and length of the scratch, various instruments have been employed, including circular punches, blades, and needles. An- tigen is applied to the abraded skin and rubbed into the scratch. The degree of erythema and wheal formation is best measured in millimeters of reaction for greater accuracy. The results are generally read in 15 to 20 minutes but the histamine controls should be measured after 10 min- utes and remeasured after 15 to 20 minutes.

The prick skin test is a variation of the puncture test. A drop of antigen is applied to the test site. The point of a needle is passed obliquely through the drop of antigen into the epidermis. The skin is gently lifted by the needle as it is withdrawn. Care is taken not to cause bleeding. The technique is simple and requires minimal technical skill. The reac- tions are generally smaller than for the scratch tests and the results are read in a similar manner and time.

Pepys and Davies (40) have considered all positive prick reactions to be significant. Others, including Aas and Belin (4), have established more rigid requirements for determining a positive response. They com- pared all prick skin tests to a histamine control (1 mg/ml). This material will generally give a 6- to 8-mm wheal when tested by the prick method and is given a value of 1 histamine equivalent. Aas has considered this to be equal to a 3+ reaction. For a test to be significant Aas requires the reaction to be equal to or greater than the histamine control.

Scratch testing enjoyed great popularity for years. It has essentially been replaced by the prick test, which is presently considered by most to be the skin test of choice. To improve reproducibility and to aid in standardization, some have recommended use of a special needle for performing the test. When prick skin testing is negative, intradermal testing with dilute antigen is recommended. Some allergists, especially those from Scandinavia, do not place much importance on allergens that react only at the intradermal level; they prefer to use immunotherapy only in those with significant positive prick skin test reactions.

The advantages of the scratch and prick skin tests are similar, and include the ability to do multiple tests easily and in a very cost-efficient manner. These tests carry a minimal risk of systemic reactions when compared to those of intradermal testing and are most easily performed, especially in young children. They require minimal equipment and much less experience and expertise than that needed for intradermal testing.

The prick skin test is preferred by many because it is more specific, sensitive, and reproducible. There is a better separation of clinically sensitive patients from normal controls. Irritant reactions are also less common with the prick skin test. Even under controlled conditions, using the same tester, the variability of reproducibility may approach 30%.

The major disadvantage of the scratch test is the lack of adequate control of the length and depth of the abrasion. The longer or deeper the cut, the greater the quantity of antigen absorbed, and the larger the reaction. The most obvious technical problem with the prick skin test is pricking through the drop of antigen.

False positive reactions for scratch or prick skin tests may be due to use of too concentrated an extract, irritants in the extract, physical trauma in performing the tests, dermatographism, reaction to preservatives in the extracts, improper interpretation, and improperly prepared or outdated antigen. False negative reactions generally are due to use of too dilute a concentration for testing, loss of potency, outdated extracts, testing technique, effects of simultaneous administration of antihistamines, or improper interpretation.

Intradermal Test. To perform this test, a 1-ml tuberculin syringe with a 26-gauge needle is used to introduce 0.02 ml of the appropriate dilution of antigen intradermally into the upper arm or forearm. Trauma should be minimized and a visible bleb should be apparent. The reactions are determined after 10 to 15 minutes using appropriate positive and negative controls. Both wheal and erythema are measured and recorded in millimeters of reaction. Various grading systems have been developed but differ according to allergist, making comparisons difficult. Many allergists resort to intradermal testing only if the scratch or prick skin test is negative. Others favor intradermal testing but test with various concentrations and generally start with more dilute solutions.

Intradermal testing is technically more difficult to perform, time-consuming, and expensive. Nonglycerinated aqueous extracts are not as stable as those with glycerin, and lower concentrations need to be prepared at approximately 7- to 14-day intervals. Fewer tests can be performed per session, and the risks of a systemic reaction is greater than for the scratch or prick skin test.

End Point Titration. Prick or intradermal titration studies can be performed for end point titration. Prick-testing antigen can be diluted in a 50% glycerin-buffered saline solution. Extracts for intradermal titration should be diluted in a buffered extraction solution containing human serum albumin as a stabilizer. Disposable needles and syringes are recommended to eliminate problems of contamination from previous allergen extracts. The back is preferred for prick testing, with the arm being ideal for the intradermal test. The end point is generally defined as the weakest concentration that will give a positive reaction (usually a 10-mm wheal or a reaction equal to the positive histamine control).

Currently, end point titration is the most precise method of quantitating the atopic skin test response. Although it has not been widely used in the past, results of several recent studies indicate a good correlation with other tests, especially inhalation challenges.

The antigen dilutions used for intradermal titration must be prepared every 7 to 14 days and kept refrigerated to minimize deterioration of potency. In addition to the technical problems inherent in doing the intradermal test, other major disadvantages include inconvenience to the patient, extra cost of equipment, and nursing time.

Inhalation Challenge Studies

Anecdotal information regarding reactions to deliberate challenges to mould material can be found in the literature of the eighteenth and nineteenth centuries. Blackley (12) reported on his own bronchospastic responses following inhalation of both *Chaetomium* and *Penicillium*.

During the early part of the twentieth century, mould sensitivity was equated to skin test reactivity on testing. It soon became apparent that many patients with skin test reactivity were not experiencing typical atopic reactions during high mould periods. Feinberg (20) proposed a number of historical factors that he felt correlated better with true mould sensitivity. For patients living in the East or Midwest, these factors would include occurrence of rhinitis and/or asthma during the summer, with no significant reactions on testing to pollen. Failure of pollen-sensitive patients to respond to immunotherapy was also felt to correlate with concomitant mould sensitivity. These patients were not symptom-free during the low-pollen season in July and early August, and continued to have symptoms past the early frost period. They generally experienced increased symptoms in dry, hot, windy weather preceded by rain, in addition to significant reactions after entering environments with a high mould content, including barns, haylofts, musty damp rooms, and basements. Symptoms could also be provoked by threshing, hay rides, or a trip to the circus.

Two of the earliest scientific attempts at correlating skin test reactions with controlled challenges were those of Harris (24) and Pennington (39). Although these studies were reported some 40 years ago, they still illustrate many of the problems associated with this type of study.

Harris (24) has reported on the results of inhalation challenges performed on 22 patients with positive skin test reactions to *Alternaria*. Twelve of these patients had a "typical history" of mould sensitivity but the remainder had a history considered doubtful by this criteria. Each patient was placed in a room into which Harris had introduced 1 g of powdered crude *Alternaria*/700 ft^3. The powdered *Alternaria* was kept circulating by the use of fans. Patients were confined to the room until symptoms developed, or for a maximum period of 1 hour. Eight patients with positive skin test reactions and a history of mould sensitivity developed asthma and/or rhinitis in from 10 to 60 minutes. Two others developed delayed symptoms starting 2 to 3 hours later, with symptoms persisting throughout the night. One patient with a positive skin test but a doubtful history also had a positive room challenge. Similar responses were noted after a second challenge by nasal insufflation of dried crude *Alternaria* material.

Pennington has reported results of skin testing and provocation studies in a group of 526 patients (39). Cotton swabs used for nasal challenge were prepared by rolling them over the surface of mould colonies grown on agar. Nasal challenges were either performed by applying these mould-laden swabs onto the nasal membranes for a maximum of 15 minutes, or by spraying aqueous extracts of these moulds directly into the nasal passage. Of 61 patients studied with positive skin tests to the mould, 22 reacted after challenge with either the swab containing the spores or with the extract: 12 patients reacted to *Hormodendrum (Cladosporium)*, 5 to *Mucor*, 4 to *Alternaria*, 2 each to *Monilia sitophilia* and *Cephalothecium roseum (Trichothecium roseum)*, and 1 each to *Aspergillus fumigatus, A. achrochus,* and *Penicillium rubrum*. Four patients developed asthma: one immediately, one after 1 to 2 hours, and the other two later that day. Systemic reactions to either skin testing or to mould immunotherapy injections were reported in 32 of these 526 patients (7.3%); 21 of these 32 patients developed symptoms with skin testing, with 6 having asthma within minutes of the skin test, and 6 of the 21 also developing symptoms after immunotherapy injections. An additional 11 patients reacted only to immunotherapy injections. Of those developing asthma, three began having symptoms 1 to 2 hours after the injection. One other patient had repeated attacks the night of the injection. Seven other patients experienced symptoms of allergic rhinitis 1 to 4 hours after the injection.

Although these and other early studies using nasal and/or inhalant provocation lacked a refined technique, they did call attention to several problems. Immediate and nonimmediate reactions were clearly identified following these early challenges, with some delayed reactions occurring the night of the challenge. There were significant numbers of false positive skin tests. In addition, Blumstein (24) reported some patients having positive reactions on challenges who had negative reactions on skin testing. This has subsequently been confirmed by others. Aas (2) has reported that 30% of patients with a positive inhalation challenge reaction to house dust had negative skin tests to the antigen. He also reported that 11% of pollen-sensitive patients react by inhalation challenges but do not react by skin testing. Two possible explanations for these discrepancies are the absence or deterioration of important antigens from the extract tested, and false positive provocation tests because of excessively irritable airways.

Inhalation challenge studies are used primarily to clarify the role of a particular allergen as the cause of allergic respiratory symptoms and to evaluate specific immunotherapy. These studies are also useful for evaluating the reliability of other *in vivo* and *in vitro* techniques, including skin testing, RAST, ELISA, and histamine release. Challenges can be used to clarify the importance of a specific exposure for a particular patient and to follow the natural course of sensitivity. On occasion they may be helpful in detecting "local sensitivity" in the face of other negative *in vivo* and *in vitro* tests.

Nasal Provocation Testing. Crude attempts at nasal challenges began in the late nineteenth century. Initial noncontrol studies introduced the antigen by blowing powdered allergen through a cone, insufflating allergen from the blunt end of a toothpick, spraying aqueous extracts with an atomizer, or placing a cotton swab rolled over the surface of a culture or soaked with aqueous extract into the nasal passage. One of the first attempts at scientific measurement of nasal airway responses to provocation testing was made by Aschan and Drettner (7) in 1958. Subsequent refinement in techniques led to the development of vaious monitoring devices and the evolution of two different types of rhinomanometry. With posterior rhinomanometry a pressure detector is placed in the mouth. Approximately 30 to 40% of patients have too sensitive a swallowing reflex and do not tolerate this procedure. For anterior rhinomanometry the detector is positioned in the unchallenged nostril. Both techniques require accurate and reproducible measurements of both pressure and flow.

Generally there is no clear relationship between the dose of antigen required to elicit a positive nasal provocation and the response to skin testing. Most workers start provocation tests at the weakest prick dose to give a positive response. A positive reaction has been variously defined as changes in the physical appearance of the nasal membranes, degree of obstruction, frequency of sneezing, nasal itching, or objective measurement of the air flow and pressure changes in the nasal passages.

The material used for the challenges can be introduced in natural or extract form. Okuda (37) has favored application of a paper disc previously impregnated with antigen and allowed to dry. One or more of these discs with known antigen content (250 μg allergen/disc) are applied to the anterior part of the inferior turbinates. Several materials have been used for negative controls, including normal saline solution, extraction fluid, and other allergens to which the patient is known to be nonreactive. The positive control is generally histamine.

Nasal provocation tests are generally regarded as research tools for evaluating the effects of immunotherapy, correlating results of other *in vivo* or *in vitro* techniques, identifying new allergens, and substituting for bronchial challenges. Major drawbacks include the time and expense required to perform the study, the problem of frequent false positive reactions, and potential systemic reactions following nasal challenge. The end point of the test needs to be more clearly defined and the procedure should be standardized. Other difficulties encountered with the use of nasal provocation tests include the inability of some patients to cooperate fully with instrument placement and technical problems relating to the equipment itself.

Explanations for discrepancies between nasal challenges and other *in vivo* and *in vitro* assessments of atopic disease are related to possible blocking antibodies in the tissue, local tissue reaginic reactivity ("latent allergy"), poorly defined end points for determining a positive response, use of an antigen of variable potency and stability, and difficulty in determining the actual antigen dose delivered. Clement et al. (14) have reported that the most significant drawback to the use of rhinomanometry is the occurrence of a nasal reflex in up to 24% of challenge patients, which leads to obstruction of the contralateral nasal passage. Using modified passive anterior rhinomanometry they were able to minimize the nasal reflex, and only 3% of the nasal provocation tests could not be reproduced.

Collins-Williams et al. (16) reported on nasal provocation testing with moulds in 150 children with perennial rhinitis. Of those tested, 92 (61%) had histories compatible with mould sensitivity. Nasal provocations were performed by having the patients inhale powdered crude extracts of the mould from the end of a toothpick. Patients were tested to a different mould every 10 minutes with the nostrils alternated, so that at least 20 minutes elapsed between challenges in the same nostril. A positive response was defined by anterior rhinoscopy and consisted of increased nasal secretions, obstruction of the nasal airway, edema of the mucosa, and subjective complaints, such as itching of the nose or sneezing. Children from 4 to 18 years of age were studied, using nine different moulds (Table 10–3). Only 11.3% of the children reacted to any mould extract by scratch testing, with 68% reacting to at least one mould extract by intradermal testing and 70.7%, reacting to at least one mould extract by

TABLE 10–3. **Comparison of Skin Test Reactions and Nasal Provocation Studies to Mould Extracts in Children***

| Allergen | Percentage Among Patients Tested | | |
| | Skin Test | | |
	Scratch	Intradermal	Provocation
Alternaria	4.5	56	24.6
Hormodendrum†	6.6	39	24.6
Aspergillus	3.6	40	10.7
Penicillium	1.3	38	18
Pullularia‡	1.3	39	19.3
Helminthosporium	2.0	50	32
Mucor	2.6	33	20
Rhizopus	1.3	26	9.3
Fusarium	2.6	36	13.3

*(Adapted from Collins-Williams et al. (16).)
†Current terminology *Cladosporium*
‡Current terminology *Aureobasidium*

nasal provocation. Positive reactions were noted in 32% of the nasal challenges to *Helminthosporium,* and 24.6% for both *Alternaria* and *Cladosporium (Hormodendrum).* Responses to the other challenges are noted in Table 10–3. Of 17 patients, 12 (70.5%) with positive scratch reactions had positive responses following nasal challenges. Of 102 patients, 69 (67.6%) with positive intradermal reactions had a positive response. Of the 31 patients with negative scratch and intradermal reactions, 9 patients (29%) gave positive responses to provocative testing. Pulmonary function testing was not performed as part of the protocol, but one of the patients developed moderately severe asthma following the challenge with *Pullularia (Aureobasidium).*

Bronchial Provocation Testing. For bronchoprovocation testing Chai et al. (13) recommended use of a dosimeter calculated to deliver a metered dose for each inspiratory capacity breath. Cockcroft et al. (15) and others have favored use of tidal breathing for a defined period, generally for 2 minutes. When these two methods were compared by Ryan et al. (44) it was shown that the same dose was delivered to the lungs regardless of the method used. With the dosimeter method, however, more aerosol was delivered to the throat and to the central airways.

The patient's subjective response and changes in auscultatory findings were the criteria for positive response in many of the early bronchial provocation studies. More definite end points have been defined, however, with an increase in our understanding of pulmonary functions and with the development of more sophisticated instruments. Using standard pulmonary function tests, a drop of 20% or more from baseline for

FEV₁ (forced expiratory volume at 1 second), *FEF* 25 to 75% (forced midexpiratory flow between 25 and 75% of the forced vital capacity), and/or *MMEFR* (midmaximal expiratory flow rate L/min) is considered to be a positive reaction. Using the body plethysmograph, an increase in airway resistance R_{aw} of greater than 50% of the initial level or a fall of 35% or more in specific airway conductance SG_{aw} is considered to be a positive response.

Evaluation of the delayed or dual reaction is somewhat more difficult. Pepys and Hutchcroft (41) have reported three different types of delayed reactions. The exact immunologic mechanism involved in these reactions is unclear. Orie et al. (38) originally proposed an association between low-dose allergen challenges and delayed reactions, but Spector and Farr (52) have recently reported delayed reactions more likely to occur in patients with high levels of IgE and at higher challenge doses.

Within the past decade a major attempt has been made to standardize the bronchial provocation test. It is generally agreed that bronchial challenges should be performed only in patients with stable asthma and only while on minimal medications (Table 10–4). There is some question in regard to testing patients while they are on steroids. Exposure to irritants, especially to ozone and nitrites, should be minimized because of the known effect of these agents on increasing bronchomotor tone. The patient's emotional state and especially recent respiratory infections are also known to affect the results of bronchial challenges.

Antigens used should be stable and nonirritating. The nebulizer used to deliver the antigen should produce particles of uniform size, approximately 1 to 3 μm. The dose delivered should be accurately measured and reproducible. The initial dose used for challenges is usually the lowest dose necessary to give a positive intradermal reaction; this is generally increased by fivefold steps until a positive response is noted or until a predetermined maximum dose is reached. The doses used for the challenge and the mould tested should be relevant to the patient's exposure.

Spector and Farr (52) have reported on the correlation of intradermal skin tests and bronchial provocation testing to various antigens, includ-

TABLE 10–4. Medications to be Withheld Prior to Bronchial Provocation Tests

Medication	Hours
Anticholinergics	8
Short-acting theophylline and β₂-agonists	8–12
Sustained-released theophyllines	24
Cromolyn	24
Most antihistamines	48
Hydroxyzine	96

ing a mould mix. They compared inhalation of a given volume of antigen in 42 patients using the Bell and Gossett pump with a second challenge using a dosimeter. A positive reaction was defined as a decrease of 20% or more in the FEV_1. If the patient had a positive intradermal reaction to a concentration of 10^{-5}, the bronchial challenge was positive 67% of the time. Positive bronchial challenges decreased to 40% and 33% with positive intradermal reactions to concentrations of 10^{-4} and 10^{-3}, respectively.

Klaustermeyer et al. (30) reported on five patients with positive intradermal reaction to several moulds and their response to subsequent inhalation challenges. They used a body plethysmograph and defined a positive challenge as a decrease of 35% in SG_{aw}, a 20% decrease in V_{max} (maximal expiratory flow rate from the peak of the MEFV curve) or a 25% decrease in $V_{max\ 50}$ (flow at 50% vital capacity from the maximal expiratory flow volume curve). Skin test reactions were noted to *Candida*, *Aspergillus*, and mould mix. Of six challenges, only three were positive: two to *Candida* and one to *Aspergillus*. It was concluded that positive responses to bronchial challenges were more likely to occur in patients who reacted to skin testing with more dilute antigen and in whom greater reactivity was observed to methacholine challenges.

In Vitro Testing Procedures

Radioallergosorbent Test (RAST)

The RAST was originally introduced by Wide et al. (56) for the *in vitro* measurement of allergen-specific IgE. Allergens are generally coupled to an insoluble carbohydrate matrix (e.g., cellulose) after activation of the matrix by cyanogen bromide. This activation results in formation of reactive groups on the matrix, which then bind covalently to amino groups on the allergens. The patient's serum is added to the insoluble matrix-allergen complex and incubated. Unreactive patient serum proteins are removed by washing. Radiolabeled ^{125}I-anti-human IgE is added and allowed to react with the IgE bound to the insoluble matrix-allergen complex. The excess of ^{125}I-labeled anti-human IgE is removed by washing and the remaining bound ^{125}I is measured in a gamma counter.

The RAST procedure may be difficult to correlate with other *in vivo* and *in vitro* tests, and this has been especially true when evaluating mould sensitivity. Coupling of some mould allergens to the insoluble matrix may be limited because of their significant glycoprotein, glycopeptide, and polysaccharide nature. Glycoprotein and glycopeptide antigens fix only sparingly because of the relatively small numbers of amino groups present. Polysaccharides lacking amino groups do not fix well to the matrix, although modified coupling procedures have been used to improve fixing. Two problems are inherent in all RAST studies regardless of the antigen being evaluated.

First, there is competitive inhibition from other classes of immunoglobulins, especially IgG. Interference by other immunoglobulins in the RAST is especially evident in patients who have been previously treated with immunotherapy (60). The excess IgG present attaches to the allergen-matrix complex and physically blocks access of specific IgE to adjacent allergen sites. This problem was believed responsible for lack of significant correlation between bronchial challenges, skin tests and RAST reported by Lynch (35) in a group of *Alternaria*-sensitive patients. He substituted Sepharose 2B for the cellulose solid phase matrix and was able to overcome this interference, essentially by increasing the number of allergen molecules attached to the matrix. IgG antibody interference has been shown to cause under-estimation of IgE antibody levels but has not been shown to convert a positive test to negative.

Secondly, RAST cannot discriminate between monovalent and polyvalent allergens. This refers to the fact that monovalent allergens or fragments do not release histamine and therefore escape detection by skin test, provocation test, or histamine release test, but are detected by RAST.

Various materials can be used for the insoluble matrix. Cellulose paper discs have been employed by Wide et al. (56). To increase binding of *Alternaria* antigens to an insoluble matrix, Yunginger et al. substituted microcrystalline cellulose for the paper discs (59). To enhance antigen binding to the matrix further and to reduce interference from IgG, workers have substituted other materials, including Agarose beads. To conserve reagents used for the RAST, Gleich et al. developed the mini-RAST (22). By coupling allergens to microcrystalline cellulose they were able to reduce the amount of allergen used by tenfold and the amount of anti-IgE by fourfold. The paper disc RAST, standard (maxi-) RAST, and mini-RAST were compared using *Alternaria tenuis* and rye grass pollen as the allergens. Microcrystalline cellulose was used as the support matrix for both the maxi- and mini-RAST. The authors concluded that there was no loss of ability to detect specific IgE and, in fact, the sensitivity was increased, thus allowing for better separation of normal controls from patients having low but significant levels of specific IgE. The mini-RAST is, however, more susceptible to interference by IgG antibodies present in patients previously treated with immunotherapy. The authors recommended use of anti-IgE of higher specific radioactivity to achieve adequate counts and better discrimination between positive reactors and controls.

Additional variations in the RAST procedure have been developed, including RAST inhibition and more recently RAST interference (21). Most of these have been reviewed by Aalberse (1).

The original Phadebas RAST marketed by Pharmacia Diagnostics lacked clinical sensitivity and specificity. It has been replaced by an updated Phadebas RAST that uses a new ^{125}I-labeled anti-IgE antibody with lower nonspecific background binding than the conventional RAST antibody, thereby permitting lowering of the RAST positive-negative cutoff. DeFelippi et al. (17) studied a panel of 13 antigens including *Alternaria* using the earlier Phadebas RAST and the updated Phadebas RAST. They demonstrated an increase in clinical sensitivity from 78 to 88% and an increase in clinical specificity from 96 to 97% using the updated Phadebas RAST. The overall clinical efficiency increased from 86 to 92%. Eight patients had a negative response to the original Phadebas RAST but gave positive responses to nasal challenges, as compared to three patients negative to the updated Phadebas RAST showing positive nasal challenges. Three patients had positive reactions to the original Phadebas RAST but negative reactions on nasal challenge, as compared to one patient who was positive to the updated Phadebas RAST but who did not respond to the nasal challenge. The authors concluded that the updated Phadebas RAST has a level of sensitivity closely approximating that of nasal provocation and skin testing.

Nalebuff et al. proposed a modification in RAST scoring to increase its sensitivity (36). Critical evaluation of this system, however, suggests that it results in false positives (25,45). Lack of standardization of the commercial RAST is a current shortcoming. Pharmacia Diagnostics uses a birch pollen reference serum and sorbent as its only standard for their RAST assays. Ideally, a separate reference system should be available for each of the antigens assayed. This would allow for more accurate assessment of RAST scores, skin test reactions, and other *in vivo* and *in vitro* studies.

Numerous studies to evaluate reaginic sensitivity have been performed using the RAST technique. Many early studies tested various allergens including moulds but usually failed to separate responses to the individual allergens. Wide et al. (56) reported a 68% agreement between results of 140 intracutaneous skin tests and RAST using 14 different antigens. Aas and Johansson (5) reported that clinical correlation between the history of sensitivity to an allergen and a positive RAST varied from 59 to 100%. The correlation improved when more defined (purified) antigens were used for testing. Berg et al. (11) studied a group of 96 children, and noted an overall agreement of RAST and provocation tests of 74%. Unfortunately, the blood samples for the RAST were not obtained at the time of the challenge and the same antigen was not used for the RAST and provocative tests. Hoffman and Haddad (27) compared a history of sensitivity with skin tests and RAST in a group of children using 13 inhalants and 14 foods. Only one mould was evaluated *(Alternaria)*. Skin test reactions to *Alternaria* were seen in 65% of the patients and 60% had positive RAST. Correlation of *Alternaria* RAST and skin test in this group was 98%. Akiyama et al. (6) reported data on end point titrations and RAST in a group of patients reactive to *Candida albicans*. Agreement between the RAST and skin test was 63%.

Aas et al. (3) evaluated 80 patients using extracts of five common moulds prepared by three different manufacturers. Skin prick tests were compared with conventional Phadebas RAST. Special RAST discs were also prepared from a partially purified extract of *Cladosporium herbarum* (Spectralgen quality) and compared to the Phadebas RAST disc. The results of these studies show discrepancies in results of skin tests and RAST, indicating a need for improvement in mould extract standardization and in RAST testing materials (Tables 10–5). These discordant findings may in part be due to use of different mould species for skin testing and RAST (Table 10–6).

Enzyme-Linked Immunosorbent Assay (ELISA)

The enzyme-linked immunosorbent assay was introduced by Engvall and Pearlmann (18) in 1972 for the detection of specific antibodies. Various modifications of this assay have been made (57) that allow for the detection of various antibodies and antigens, including haptens. The classic indirect ELISA and the double antibody techniques (47) are compared in Figure 10–2.

Single Antibody (Indirect) Method. The antigen to be tested is ab-

TABLE 10–5. Comparison of Phadebas RAST Reactions with Skin Test Reactions Using Three Different Commercial Antigens*

Mould	Percentage Among Patients Tested			
	Phadebas RAST	Extract C	Extract D	Extract F
Alternaria	59	10	14	35
Aspergillus	62.5	10		42.5
Cladosporium herbarum	67.5	12.5	30	65
Mucor	40	1	9	27.5
Penicillium	56	2.5	9	32.5

*(Modified from Aas et al. (3). Permission obtained from Munksgaard International Publishers, Copenhagen.)

TABLE 10–6. Composition of Phadebas RAST and Three Commercial Extracts

Mould	Phadebas RAST	Extract C	Extract D	Extract F
Alternaria	*A. tenuis*	*A. iridis*	*A. tenuis*	*A. tenuis*
Aspergillus	*A. fumigatus*	*A. fumigatus*	*A. spinosus*	*Aspergillus* mix*
Cladosporium	*C. herbarum*	*C. herbarum*	*C. herbarum*	*C. herbarum*
Mucor	*M. racemosus*	*M. racemosus*	*M. mucedo*	*M. racemosus*
Penicillium	*P. notatum*	*P. roqueforte*	*P. notatum*	*Penicillium* mix†

Aspergillus mix contained *A. fumigatus, A. terreus, A. niger,* and *A. nidulans.*
†*Penicillium* mix contained *P. digitatum, P. expansum, P. glaucum, P. roseum,* and *P. notatum.*
(Adapted from Aas et al. (3). Permission obtained from Munksgaard International Publishers, Copenhagen.)

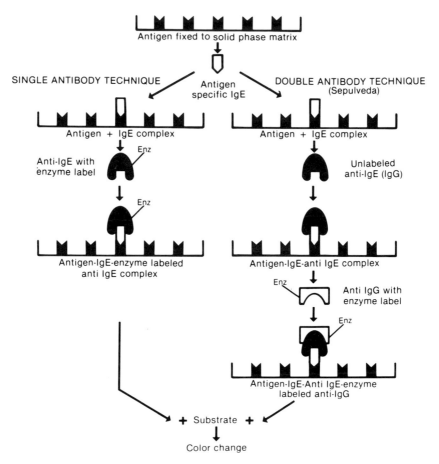

Fig. 10–2. **Comparison of single antibody and double antibody technique for the ELISA.**

sorbed onto a solid phase matrix or is chemically bound to a solid phase matrix. For optimal results the antigen should be in moderate excess. After washing, the patient's serum is added, and immunoglobulins present combine with the solid phase matrix-antigen complex. Unattached immunoglobulins are washed off and an enzyme-labeled anti-IgE is added. Several hours are generally allowed for maximum coupling of the anti-IgE to the solid phase matrix-antigen complex to occur. The anti-IgE that does not couple is washed from the system. The enzyme marker reacts with an appropriate substrate, and the color change in the substrate is proportional to the IgE present in the patient's serum, as quantitated by spectrophotometric assay.

Double Antibody (Indirect) Method. The initial stages of this assay are identical to those for the single antibody method. An unlabeled rabbit anti-IgE is substituted for the enzyme-labeled anti-IgE in the single antibody system. After allowing sufficient time for coupling of the anti-

IgE with the solid phase matrix-antigen complex, the excess anti-IgE is washed off the solid-phase antigen anti-IgE complex. An enzyme-labeled anti-rabbit IgG is added and is allowed to react with the rabbit anti-IgE. The assay from this point is the same as that for the single antibody technique. Sepulveda (47) compared the indirect ELISA and double antibody techniques and found the latter to be more sensitive. Other advantages of the double antibody technique are the commercial availability of sensitive antisera and the increased reactivity possible between IgE and unlabeled anti-IgE. Direct labeling of the anti-IgE leads to distortion of this antibody and diminished coupling with the carrier-antigen complex. Various enzymatic solid phase matrix substrates and methods for fixing enzyme to immunoglobulin carriers have been developed. These have been reported and reviewed by many workers (31,55,57).

ELISA was initially developed as an alternative to the RAST assay. It has the major advantage of requiring less sophisticated equipment and reagents. Less skilled personnel can administer this test, and there is no radiation hazard. The assay is relatively inexpensive yet offers a high degree of sensitivity and specificity. Protein and nonprotein antigens can be detected. The RAST is generally more sensitive and subject to less interference than the ELISA. Disadvantages of the ELISA method include greater difficulty in defining the end point of the reaction, greater interference from icteric and hemolyzed serum, nonspecific background enzyme activity, and difficulty in labeling and purifying the enzyme conjugate. The RAST and ELISA are compared in Table 10–7.

Eriksson and Ahlstedt (19) compared the skin test, nasal and bronchial challenge tests, clinical history, ELISA, and RAST. They studied only house dust and several pollen and animal dander. The authors prepared their own reagents for the ELISA using alkaline phosphatase conjugated to antihuman IgE. The RAST studies were performed using the commercially available reagents from Pharmacia. A negative ELISA had a high probability of indicating nonallergy and a very high ELISA generally indicated allergy. In most tests, however, the ELISA was doubtful or discordant with the RAST, skin tests, and results of the challenge studies. The authors concluded that the ELISA, as performed in this study, offered no advantage over other methods for diagnosing atopic disease.

Histamine Release From Basophils

Histamine release occurs in a dose-dependent fashion from sensitized basophils when exposed to appropriate antigen in the presence of Ca^{++}. The release of preformed histamine is an active cellular process modulated by intracellular cyclic adenosine monophosphate (cAMP). Histamine release is, to a major degree, influenced by medications that act on cAMP production. Beta-agonist and theophylline medications must be avoided for a minimum of 24 to 48 hours before blood samples are obtained. The amount of histamine released in the assay is dependent

TABLE 10–7. **Comparison of RAST and ELISA for Detection of Mould Reaginic Sensitivity**

	RAST	ELISA
Antigen	Protein Polysaccharides (modified coupling) Glycoprotein Glycopeptides	Protein Polysaccharides Glycoprotein
Carrier (insoluble)	Carbohydrate matrix Sephadex Agarose Cellulose (microcrystalline) Paper disc Walls of polystyrene test tubes	Carbohydrate matrix Sephadex Agarose Polyacrylamide Paper disc Absorption onto polystyrene, glass slides, and cellulose-ester filter disc (Millipore)
Label for Detection of IgE	^{125}I-labeled anti-IgE measured in gamma counter	Enzyme-labeled anti-IgE measured spectrophotometrically
Advantages	Specific and sensitive assay; generally more sensitive than ELISA Minimal interference problems, not affected by hemolysis or icteric serum End point of reaction easier to define Semiautomated	Test easy to perform and rapid with potential for automation Can be performed by less skilled personnel Low risk; no radioactivity Specific and sensitive Reagents relatively inexpensive and readily available Reagents have somewhat longer shelf life Various enzyme labels available Availability of commercial antihuman IgE of good potency
Disadvantages	Equipment is costly Nonspecific interference when IgE is elevated Enzymes or enzyme inhibitors present in spore-rich mold extracts may uncouple antigen, interfere with coupling process, inactivate or alter antigen Requires skilled and specially licensed personnel Strict regulations for handling radioactive material Reagents potentially hazardous and costly, with relatively short shelf life	Equipment is costly Nonspecific interference when IgE is elevated Enzymes or enzyme inhibitors present in spore-rich mold extracts may uncouple antigen, interfere with coupling process, inactivate or alter antigen Presence of background enzyme Enzyme used in assay cannot also be present in antigen; must also exclude enzyme inhibitors and substrate Expertise in labeling and conjugate purification required Distortion and possible diminished activity of enzyme carrier as a result of enzyme labeling End point of reaction is harder to define Not as sensitive as RAST Prone to nonspecific interference from icteric or hemolyzed serum
Sensitivity	Semiquantitative	Semiquantitative
Commercial Availability	Phadebas RAST (Pharmacia) Kallested RAST	Phadezym RAST (Pharmacia)
Antigens Available (Mould)	*Alternaria tenuis* *Aspergillus fumigatus* *Candida albicans* *Cladosporium herbarum* *Mucor racemosus* *Penicillium notatum*	*A. tenuis* *A. fumigatus* *C. albicans* *C. herbarum* *M. racemosus* *P. notatum*

on the amount of specific IgE and IgG (blocking antibody) present in addition to the ratio of these immunoglobulins and the intrinsic ability of the basophils present to release histamine.

Bioassay methods were initially used by Barsoum and Gaddum (10) for histamine determinations. In 1959 Shore et al. (48) described a rather cumbersome fluorimetric method that has subsequently been modified to improve sensitivity and reproducibility. Automation has further improved its accuracy and sensitivity (49). More recently Snyder et al. (51) described an enzyme-isotope method and Skov et al. (50) used a radioactive assay. The automated fluorimetric technique is presently the most widely used method for histamine assay.

Fluorimetric Technique. In this method protein is removed from the serum by precipitation with percholoric acid. The histamine content is extracted in *n*-butanol to remove histadine and other potentially interfering materials. The histamine is coupled in a dilute HCl solution with *o*-phthalaldehyde to produce a fluorescent histamine-*o*-phthalaldehyde complex that can be measured in a fluorometer.The manual method for fluorimetric assay of histamine requires approximately 1 ml blood/determination. In the automated system approximately 30 samples can be assayed per hour using a sample of 0.5 ml of blood. It is possible to assay from 0.10 to 15 mg histamine with a precision of 1 to 2%. Although the automated system is more expensive to set up, it is more cost-efficient if large numbers of samples are handled.

Enzymatic-Isotopic Technique. This assay (50) is dependent on the methylation of histamine by histamine methyltransferase using a ^{14}C-methyl histamine. Trace amounts of ^{3}H-histamine are added to the system, and labeled and unlabeled histamine are determined. The enzyme-isotope assay is as sensitive as the fluorimetric method but has the advantages of being more reproducible, more easily automated, and less time-consuming. By this method histamine levels can be measured accurately within the range of 0.1 to 2 ng.

Radioactive Label Technique. Sensitized basophils are incubated with ^{3}H, which incorporates the label into the intracellular histamine. When challenged with appropriate antigen, radiolabeled ^{3}H-histamine is released along with unlabeled histamine. The percentage of histamine released is determined by comparing the amount of ^{3}H-histamine in the supernatant with the total radiolabeled histamine present (50).

A significant problem with histamine release assays is the frequent false positive and negative reactions encountered. Using a highly purified venom system, Kagey-Sobotka et al. (28) reported a 15 to 17% false positive rate with a false negative rate of 19%. Similar comparative studies are not available for mould antigens. Additional problems encountered with histamine release include the spontaneous histamine release seen in some patients, the natural fluorescence of cellular toxins in the antigen being assayed, and the presence of blocking antibodies from previous immunotherapy.

Most work with histamine release has been accomplished using washed patient leukocytes, although some researchers prefer to use whole blood and others use passive sensitization of donor cells. The latter approach allows for quantitation of relative amounts of specific IgE antibody and for more direct comparison to the RAST. The average amount of histamine present in blood is 83 ng/ml. The results of these assays are expressed as percentages of histamine release/total histamine present in the blood sample, with a correction made for the histamine released spontaneously (generally 1–5%). Various end points have been defined, which generally range from 20 to 50% total histamine present. Use of anti-IgE is recommended as a positive control to ensure that adequate histamine can be released from the basophils present. This would preclude interference from medications.

Reinders et al. (43) have reported results of histamine release in a group of *Alternaria*-sensitive patients, some of whom had been on immunotherapy. Center *Alternaria* was used for the histamine release and skin tests and compared with results of Pharmacia RAST. Patients treated with immunotherapy had received Center *Alternaria* extract. Shore's fluorimetric assay was used to measure histamine release. Of the 18 patients with positive skin tests, 4 were on immunotherapy, receiving 500 PNU of *Alternaria* every 3 weeks as maintenance. The treated and untreated group could not be separated by either RAST or histamine release studies. There was also no correlation between histamine release and results of RAST. An inherent difficulty with this study was the use of Center *Alternaria* for skin tests and histamine release with comparisons made to RAST using Pharmacia *Alternaria* antigen. Some discrepancies of this study undoubtedly reflect major qualitative differences in the two *Alternaria* antigens.

COMPARISON OF DIAGNOSTIC PROCEDURES IN MOULD ALLERGY

Tests used for confirmation of mould allergy should be standardized, accurate, and reproducible, inexpensive, and easy to perform, with minimal risk for the patient. No test presently available fulfills all these requirements. Table 10-8 contrasts three of the most commonly used tests: skin test, histamine release from basophils, and RAST. The major advantages of skin testing include the relative ease of performance, immediate results, minimal cost, and the ability to perform multiple tests at one time. It cannot be used if significant dermatographism, atopic dermatitis, or urticaria are present; also, it carries a minimal risk for the patient to have an anaphylactic reaction. Histamine release from basophils requires a significant blood sample, and the study must be done promptly. It is relatively expensive to perform and, because of technical

TABLE 10–8.   Comparison of Three Commonly Used Tests: Skin Test, Histamine Release, and RAST

	Skin Test	Histamine Release	RAST
Amount of antigen required	Minimal	Small	Small
Quality of antigen	Purification not required; reliability and reproducibility improved with purified material	Purification not required; reliability and reproducibility improved with purified material	More accurate with purified antigen
Coupling	None	None	Required
Standardization-end point	Relatively easily defined	Depends on donor	Easy to standardize to particular material
Interference from blocking antibodies	Yes, from previous immunotherapy	Same	Same
Advantages			
Expense	Minimal	Moderate to expensive	Moderate
Possible anaphylaxis	Small but significant	None	None
Radiolabel	None	None	Yes
No. tests performed/time	Very large	Limited	Somewhat limited
Miscellaneous	Must avoid if significant skin disease; limited in very young child; no blood sample required	Some restriction based on blood sample required	Least restrictive
		Needs significant amounts of blood	Serum can be stored for analysis or comparison at later date
	End point titration probably more sensitive than RAST or histamine release	Similar to RAST	
		Sample must be analyzed promptly	

problems, only a limited number of studies can be performed at one time. RAST has been the most popular *in vitro* test, partly because it requires a small blood sample, serum can be stored for later testing, a fairly large number of antigens can be studied, and there is minimal patient discomfort.

By their nature inhalation challenges do not fulfill many of the requirements noted above. Challenge studies, for the most part, should be limited to special circumstances, such as studies of previously unrecognized allergens, evaluation of the effectiveness of immunotherapy, confirmation of sensitivity when a major discrepancy exists between the results of skin testing, history, and other *in vitro* tests, and evaluation of occupational problems.

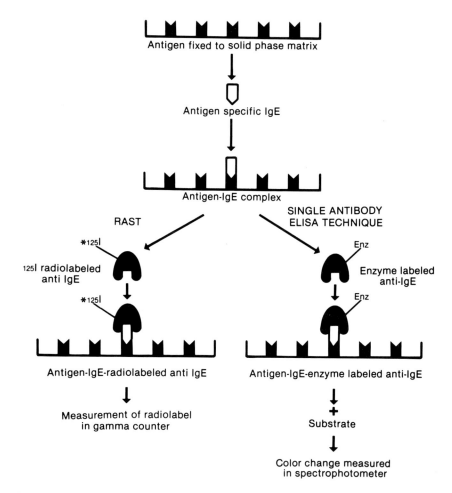

Fig. 10–3. Comparison of RAST and single antibody ELISA methods for detection of specific IgE.

Two widely promoted *in vitro* tests for determining reaginic sensitivity are RAST and ELISA. These are compared in Table 10–7 and Figure 10–3. At present RAST appears to have many advantages over ELISA. The latter has numerous problems, especially those related to the enzyme content of the various mould extracts, potential presence of substrate and enzyme inhibitors in the extract, and difficulties in labeling and in conjugate purification.

The presence of mould sensitivity can best be established by clinical history, appropriate prick and intradermal titration, with confirmation by RAST, and/or when necessary by suitable challenges. Many studies comparing the various *in vivo* and *in vitro* techniques have used extracts from different manufacturers. Because the Phadebas RAST is readily available, these antigens are commonly compared to extracts from other manufacturers. Such problems are illustrated by the results of five recent studies of mould extracts (3,6,8,32,53).

The report by Aas et al. (3) is particularly pertinent. Five mould extracts (*Alternaria, Aspergillus, Cladosporium, Mucor,* and *Penicillium*) were compared by prick skin test, RAST, and bronchial provocation studies. Extracts were obtained from three different commercial manufacturers, with two companies supplying two different extracts each. Positive prick skin tests for *Cladosporium herbarum* varied from 12.5 to 60%. The Phadebas RAST was positive in 67.5% of the patients. A reference *C. herbarum* material and a purified extract of *C. herbarum* were used for special RAST studies. The RAST was positive in 79% of patients using the former material and in 86% with the latter. Bronchial challenges were performed with only one of the extracts. Of 22 patients with a RAST score of 2+ or higher and with a 2+ or larger prick skin test reaction, 13 had a positive bronchial challenge.

Future studies should employ the same mould extract in all phases of testing, including RAST and ELISA. When one extract is compared to another, comparable materials and concentrations should be used. The extracts must contain appropriate antigens, be standarized for potency, and be properly stored to prevent deterioration.

REFERENCES

1. Aalberse, R.C.: IgE-based radioimmunoassays for the quantitation of allergens. Allergy, **35**:236, 1980.
2. Aas, K.: The Bronchial Provocation Test. Springfield,IL, Charles C Thomas, 1975.
3. Aas, K., et al.: Immediate-type hypersensitivity to common molds. Comparison of different diagnostic materials. Allergy, **35**:443, 1980.
4. Aas, K., and Belin, L.: Standardization of diagnostic work in allergy. Acta Allergol., **27**:439, 1972.
5. Aas, K., and Johansson, S.G.O.: The radioallergosorbent test in the in vitro diagnosis of multiple reaginic allergy. J. Allergy Clin. Immunol., **48**:134, 1971.
6. Akiyama, K., et al.: Relationship between the results of skin, conjunctival and bronchial tests and RAST with *Candida albicans* in patients with asthma. Clin. Allergy, **11**:343, 1981.

7. Aschan, G., and Drettner, B.: Nasal obstruction at provocation experiments in patients with hay-fever. Acta Otolaryngol. [Suppl.] (Stockh.) **140**:91, 1958.
8. Aukrust, L., and Aas, K.: The diagnosis of immediate-type allergy to *Cladosporium herbarum*. Differences between extracts and the efficacy of radioallergosorbent test (RAST) assays. Allergy, **33**:24, 1978.
9. Baer, H., and Hooton, M.L.: The effect of preservatives on the stability of short ragweed pollen extract. Div. Biol. Stand., **24**:115, 1974.
10. Barsoum, G.S., and Gaddum, J.H.: The pharmacological estimation of adenosine and histamine. J. Physiol. (Lond.), **85**:1, 1935.
11. Berg, T., Bennich, H., and Johansson, S.G.O.: In vitro diagnois of atopic allergy. I. A comparison between provocation tests and the radioallergosorbent test. Int. Arch. Appl. Allergy, **40**:770 1971.
12. Blackley, C.H.: Experimental Research on the Cause and Nature of Catarrhus Aestivus. London, Bailliere, Tindall and Cox, 1873.
13. Chai, H., et al.: Standardization of bronchial inhalation challenge procedures. J. Allergy Clin. Immunol., **56**:323,1975.
14. Clement, P.A.R., et al.: Nasal provocation and passive anterior rhinomanometry (PAR). Clin. Allergy, **11**:293, 1981.
15. Cockcroft, D.W., et al.: Bronchial reactivity to inhaled histamine; a method and clinical survey. Clin. Allergy, **7**:235, 1977.
16. Collins-Williams, C., et al.: Nasal provocative testing with molds in the diagnosis of perennial allergic rhinitis. Ann. Allergy, **30**:557, 1972.
17. DeFilippi, I., Yman, L., and Schroder, H.: Clinical accuracy of updated version of the Phadebas RAST test. Ann. Allergy, **46**:249, 1981.
18. Engvall, E., and Pearlmann, P.: Enzyme-linked immunosorbent assay, ELISA. III. Quantitation of specific antibodies by enzyme-labelled anti-immunoglobulin in antigen-coated tubes. J. Immunol., **109**:129, 1972.
19. Eriksson, N.E., and Ahlstedt, S.: Diagnosis of reaginic allergy with house dust, animal dander and pollen allergens in adult patients. V. A comparison between enzyme-linked immunosorbent assay (ELISA), provocation tests, skin tests and RAST. Int. Arch. Allergy Appl. Immunol., **54**:83, 1977.
20. Feinberg, S.M.: Seasonal hay fever and asthma due to molds. JAMA, **107**:1861, 1936.
21. Gleich, G.J., et al.: Measurement of IgG blocking antibodies by interference in the radioallergosorbent test. J. Immunol., **126**:575, 1981.
22. Gleich, G.J., Adolphson, M.S., and Yunginger, J.W.: The Mini-RAST: Comparison with other varieties of the radioallergosorbent test for the measurement of immunoglobulin E antibodies. J. Allergy Clin. Immunol., **65**:20, 1980.
23. Gleich, G.J., and Jones, R.T.: Measurement of IgE antibodies by the radioallergosorbent test. I. Technical considerations in the performance of the test. J. Allergy Clin. Immunol., **55**:334, 1975.
24. Harris, L.H.: Experimental reproduction of respiratory mold allergy. J. Allergy, **12**:279, 1941.
25. Hoffman, D.R.: Comparison of methods of performing the radioallergosorbent test: Phadebas R., Fadal Nalebuff and Hoffman protocols. Ann. Allergy, **45**:343, 1980.
26. Hoffman, D.R., et al.: Isolation of spore-specific allergens for *Alternaria*. Ann. Allergy, **46**:310, 1981.
27. Hoffman, D.R., and Haddad, A.H.: Diagnosis of multiple inhalant allergies in children by radioimmunoassay. Pediatrics, **54**:151, 1974.
28. Kagey-Sobotka, A.K., et al.: Measurement of IgG-blocking antibodies: Development and application of a radioimmunoassay. J. Immunol., **117**:84, 1976.
29. Kimura, P., Lopez, M., and Salvaggio, J.: Characterization of types of enzymatic activity in somatic extracts of selected fungi, thermophilic actinomycetes and pollen by immunoelectrophoresis. Clin. Allergy, **5**:331, 1975.
30. Klaustermeyer, W.B., et al.: The asthmatic airway response to inhaled antigen. Ann. Allergy, **45**:338, 1980.
31. Kricka, L.J., et al.: Variability in the absorption properties of microtitration plates used as solid support in enzyme immunoassay. Clin. Chem., **25**:741, 1980.
32. Kurimoto, Y., and Baba, S.: Specific IgE estimations by RAST in Japanese asthmatics compared with skin, passive transfer and bronchial provocation tests. Clin. Allergy, **8**:175, 1978.

33. Lowenstein, H.: Quantitative immunoelectrophoretic methods as a tool for the analysis and isolation of allergens. Prog. Allergy, 25:1, 1978.
34. Lowenstein, H., Aukrust, L., and Gravensen, S.: *Cladosporium herbarum* extract characterized by means of quantitative immunoelectrophoretic methods and special attention to immediate-type allergy. Int. Arch. Allergy Appl. Immunol., 55:1, 1977.
35. Lynch, N.R.: Influence of IgG antibody and glycopeptide allergens on the correlation between the radioallergosorbent test (RAST) and skin testing or bronchial challenge with *Alternaria*. Clin. Exp. Immunol., 22:35, 1975.
36. Nalebuff, D.J., Fadal, R.G., and Ali, M.: The study of IgE in the diagnosis of allergic disorders in an otolaryngology practice. Otolaryngol. Head Neck Surg., 87:351, 1979.
37. Okuda, M.: Nasal provocation. Advances in Allergy and Applied Immunology. Proceedings of the 10th Intern. Cong. Allergology, 1980.
38. Orie, N.G.M., et al.: Late reactions in bronchial asthma. In Intal in Bronchial Asthma. Edited by J. Pepys and Y. Yamamura. London, Beck and Partridge, 1974.
39. Pennington, E.S.: A study of clinical sensitivity to airborne molds. J. Allergy, 21:388, 1941.
40. Pepys, J., and Davies, R.J.: Allergy. In Asthma. Edited by T.J.H. Clark and S. Godfrey. Philadelphia, W.B. Saunders, 1977.
41. Pepys, J., and Hutchcroft, B.J.: Bronchial provocation tests in etiologic diagnosis and analysis of asthma. Am. Rev. Resp. Dis., 112:829, 1975.
42. Personal communication.
43. Reinders, E.J., et al.: Human leukocyte histamine release assays with whole ragweed and *Alternaria* antigens using the Technicon Autoanalyzer II. Allergy, 35:391, 1980.
44. Ryan, G., et al.: Standardization of inhalation provocation tests: Two techniques of aerosol generation and inhalation compared. Am. Rev. Resp. Dis., 123:195, 1981.
45. Santrach, P.J., et al.: Diagnostic and therapeutic applications of a modified radioallergosorbent test and comparison with the conventional radioallergosorbent test. J. Allergy Clin. Immunol., 67:97, 1981.
46. Schumacher, M.J., et al.: Primary interaction between antibody and components of *Alternaria*. II. Antibodies in sera from normal, allergic and immunoglobulin-deficient children. J. Allerg. Clin. Immunol., 56:54, 1975.
47. Sepulveda, R.: Measurement of IgE and IgG antibodies by ELISA technique. Advances in Allergy and Applied Immunology. Proc. 10th Intern. Cong. Allergology, 1980.
48. Shore, P.A., Burhalter, A., and Cohn, V.H.: A method for fluorometric assay of histamine in tissues. J. Pharmacol Exp. Ther., 127:182, 1959.
49. Siraganian, R.P., and Brodsky, M.J.: Automated histamine analysis for in vitro allergy testing. I. A method utilizing allergen-induced histamine release from whole blood. J. Allergy Clin. Immunol., 57:525, 1976.
50. Skov, P.S., Norm, S., and Weeke, B.: ^{3}H-histamine release from human leukocytes. A new method for detecting type I allergy compared with basophil histamine release technique. Allergy, 34:261, 1979.
51. Snyder, S.H.R., Baldessarini, R.J., and Axelrod, J.: A sensitive and specific enzymatic isotopic assay for tissue histamine. J. Pharmacol. Exp. Ther., 153:544, 1966.
52. Spector, S., and Farr, R.: Bronchial inhalation challenge with antigens. J. Allergy Clin. Immunol., 64:580, 1979.
53. Turner, K.J., et al.: Standardization of allergen extracts by inhibition of RAST, skin test and chemical composition. Clin. Allergy, 10:441, 1980.
54. Voller, A., Bidwell, D.E., and Bartlett, A.: Enzyme immunoassays in diagnostic medicine. Theory and practice. Bull. WHO, 53:55, 1976.
55. Voller, A., Bidwell, D., and Bartlett, A.: Enzyme-linked immunosorbent assay. In Manual of Clinical Immunology. 2nd Ed. Edited by N. Rose and H. Friedman. Am. Soc. Microbiol., :359, 1980.
56. Wide, L., Bennich, H., and Johansson, S.G.O.: Diagnosis of allergy by an in vitro test for allergen antibodies. Lancet, 2:1105, 1967.
57. Wisdom, G.B.: Enzyme-immunoassay. Clin. Chem., 22:1243, 1976.
58. Yunginger, J.W., Jones, R.T., and Gleich, G.J.: Studies on *Alternaria* allergens. II. Measurement of the relative potency of commercial *Alternaria* extracts by the direct RAST and by RAST inhibition. J. Allergy Clin. Immunol., 58:405, 1976.
59. Yunginger, J.W., Roberts, G.D., and Gleich, G.J.: Studies on *Alternaria* allergens. I.

Establishment of the radioallergosorbent test for measurement of *Alternaria* allergens. J. Allergy Clin. Immunol., **57**:293, 1976.

60. Zimmermann, E.M., Yunginger, J.W., and Gleich, G.J.: Interference in the ragweed pollen and honeybee venom radioallergosorbent tests. J. Allergy Clin. Immunol., **66**:386, 1980.

Chapter 11

REVIEW OF THERAPEUTIC
PROCEDURE FOR MOULD ALLERGY

Joanne F. Domson

Mould allergy can manifest itself in any phase of the spectrum of allergic disease but is most commonly recognized as respiratory allergy, either rhinitis and/or asthma. Currently there are three widely accepted methods of treating respiratory allergy: avoidance or environmental control measures, pharmacotherapy, and immunotherapy. These will be discussed in detail as they pertain to allergic respiratory disease (rhinitis and asthma) caused by fungus. Treatment of allergy to ingested fungi or possible dermatologic manifestations of allergy to fungi is controversial and has not been scientifically documented, so therefore it will not be included in this chapter.

PHARMACOTHERAPY

The pharmacotherapy of asthma and allergic rhinitis, whether it is intrinsic or secondary to extrinsic allergy, has been studied extensively during the last decade. This research has been responsible for the development of many new drugs (cromolyn, albuterol, and beclomethasone), which have expanded the clinician's treatment armamentarium and have improved prognosis for many patients.

Allergic Rhinitis

Understanding the pathophysiology of allergic rhinitis is necessary for proper pharmacotherapy but is incompletely understood at present. A recent article by Mygind from Copenhagen (24) has outlined the cur-

TABLE 11–1.   Antihistamine-Decongestant Combinations

Group	Drug
I	Rondec
	Carbinoxamine maleate
	Pseudoephedrine
II	Poly-Histine D
	Pyrilamine
	Pheniramine
	Phenylpropanolamine
	Phenyltoloxamine
III	Dimetapp
	Brompheniramine
	Phenylpropanolamine
	Phenylephrine
	Fedahist
	Chlorpheniramine
	Pseudoephedrine
	Actifed
	Triprolidine
	Pseudoephedrine
IV	Atarax, Sudafed
V	Phenergan VC
	Promethazine
	Phenylephrine

rent knowledge of the mechanisms of nasal allergy and has provided a basis for rational pharmacotherapy.

A plethora of antihistamine-decongestant drugs has flooded the market (Table 11–1); management of allergic rhinitis, however, although improved, is still difficult. In addition, the benefits of drugs must be weighed against their side-effects.

Antihistamines (or histamine H_1 receptor antagonists), as the name implies, work through competitive inhibition of histamine and are effective, if given early, in reducing many patients' sneezing, itching, and rhinorrhea. If their effectiveness tends to decrease with prolonged use, switching to a drug from a different class of antihistamines is sometimes effective (Table 11–2). Unfortunately, antihistamines may cause sedation or nervousness, dryness of the mouth, blurred vision, decreased mental acuity, and insomnia. These side-effects vary with the particular class of antihistamine and the patient's physiologic condition. Therefore, drugs from a particular class may be chosen for the most desired effect; that is, relief of itching as well as motion sickness increases with drugs from classes IV and V, while somnolence is an important side-effect of those from class I.

TABLE 11–2. Antihistamine Classes

Class No.	Class Name
I	Ethanolamines
	Diphenhydramine (Benadryl)
	Carbinoxamine (in Rondec)
II	Ethylenediamines
	Tripelennamine (Pyribenzamine, PBZ)
	Thenylpyramine (Histadyl)
III	Alkylamines
	Chlorpheniramine (Chlor-Trimeton)
	Brompheniramine (Dimetane)
	Triprolidine (Actidil)
	Dexchlorpheniramine (Polaramine)
IV	Piperazines
	Hydroxyzine (Atarax)
	Cyclizine (Marezine)
V	Phenothiazines
	Promethazine (Phenergan)
	Trimeprazine (Temaril)
	Metadelazine (Tacaryl)
VI	Miscellaneous
	Clemastine fumarate (Tavist)
	Cyproheptadine hydrochloride (Periactin)
	Azatadine (Optimine)

Oral decongestants, or alpha-adrenoceptor agonists, including ephedrine, phenylephrine, and pseudoephedrine, act by stimulating contraction of the smooth muscles of the nasal mucosal vasculature. Among side-effects, however, are elevation of blood pressure, nervousness, and restlessness. These drugs should therefore be avoided in patients with hypertension, heart disease, or hyperthyroidism.

Decongestants, when applied topically to the nose, have given relief, but with prolonged use there is the risk of developing "rhinitis medicamentosa" or "rebound phenomenon" which is characterized by a marked increase in nasal swelling and congestion shortly after the initial decongestion. This condition can apparently be reversed by using topical nasal steroids and avoiding the topical decongestant.

Aerosol steroid preparations, such as beclomethasone (Vancenase, Beconase) and flunisolide (Nasalide), have been introduced recently for topical treatment of allergic rhinitis. The recommended doses of 300 to 400 µg/day were found to produce no adrenal suppression (6,25) and provided remarkable relief in 70 to 80% of patients with extrinsic seasonal and perennial allergic rhinitis (25). These drugs were also found to reduce the size of nasal polyps in 45% of patients studied. Even after 2 years of continuous use of flunisolide for perennial rhinitis (10), patients experienced subjective benefit without serious side-effects or evidence of adrenal suppression. The consequences of using topical steroids have recently been identified as the following:

1. Reduction in the number of epithelial mediator cells.
2. Reduction of epithelial and endothelial permeability.
3. Reduction of the reflectory response to mechanical stimulation of sensory nerves.
4. Reduction of the secretory response to stimulation of glandular cholinoceptors.
5. Partial inhibition of the immediate allergen-induced nasal symptoms (24).

Systemic steroids such as prednisone, prednisolone, or decadron are only rarely recommended for a severe bout of allergic rhinitis. Because prolonged use results in adrenal suppression, these steroids should be given for short periods. For example, prednisone dosage should not exceed 40 to 60 mg per day and should be tapered over 4 to 5 days under the close supervision of a physician.

Recently, intranasal topical cromolyn sodium has been proved effective in improvement of seasonal and perennial allergic rhinitis (9,41); this drug is now available commercially (Nasalcrom). Because of the importance of reflex vasodilation and hypersecretion mediated through the parasympathetic reflex arc in the nose, however, it is required that almost 100% of the histamine release be blocked by the cromolyn for it to be effective (24).

Topical antihistamines are also being investigated for use in allergic rhinitis. Secher et al. (37) have shown that a combination of H_1 and H_2 antihistamines topically produces an additive effect on the relief of histamine-induced increase in nasal airway resistance. The H_1 antihistamine was chlorpheniramine maleate and the H_2 antihistamine was ranitidine hydrochloride.

Asthma

Asthma is a chronic disease characterized by increased responsiveness of the tracheobronchial tree to various stimuli. There is an increase in spasm of the smooth muscle surrounding large and small bronchioles, edema of the submucosa, and hypersecretion of mucous from the glands lining the airways. Nonspecific stimuli, such as weather changes, anxiety, exercise, and infection may trigger an asthmatic response, as may specific allergens such as fungi.

Beta-adrenergic drugs form the cornerstone of the pharmacotherapy of asthma. Epinephrine is still widely accepted as the drug of choice in an acute asthmatic attack. Bronchodilation is rapidly achieved by stimulation of the β_2 receptors on bronchial smooth muscle cells. This β_2

stimulation also results in decreased mediator release from mast cells. The disadvantage of sympathomimetic drugs such as epinephrine is that, in addition to β_2 stimulation, they also cause stimulation of β_1 and alpha-adrenergic receptors. The end result is cardiac stimulation and an increase in blood pressure, with the possibility of cardiac arrhythmias and fibrillation. Epinephrine and other sympathomimetic drugs also metabolize rapidly and must be given by injection.

With epinephrine and isoproterenol as the templates, new drugs with β_2 specificity have been developed. Norn and Skov have reviewed the usefulness of these selective β_2 stimulators: metaproterenol, terbutaline, and salbutamol (30). These drugs were found to have no alpha-stimulating effect and hardly any beta-stimulating effect. Therefore, there are minimal cardiac inotropic or chronotropic effects. Another advantage is that these drugs may be administered either orally, by inhalation, or by injection. They are chemically stable and are metabolized slowly, so their duration of action is up to 6 to 8 hours. The side-effects of these β_2-specific drugs include muscle tremor, slight increase in heart rate, and headache; they most often accompany oral administration of the drug.

Another drug, theophylline, was introduced early in this century, but fell into disfavor because of reports of drug-related deaths (44). The thorough and careful work of Weinberger et al. (43,45) in the last decade has reinstated theophylline to its present valuable position in the treatment of asthma. This has resulted in the development of general guidelines for dosage requirements based on the patient's age and weight. Wyatt et al. have also shown the variability of metabolism of this drug, necessitating frequent monitoring of blood levels of the drug to maintain it in a therapeutic but nontoxic range (10–20 μg/ml) (45). The usual dose that will result in these levels varies with age:

Under 9 years: 24 mg/kg/24 hr
9 to 12 years: 20 mg/kg/24 hr
12 to 16 years: 18 mg/kg/24 hr
Over 16 years: 12 to 13 mg/kg/24 hr

Theophylline, a phosphodiesterase inhibitor, acts by increasing the cAMP:cGMP ratio, resulting in dilation of bronchial smooth muscle cells, and by inhibiting mediator release from mast cells. Liquid preparations result in peak serum levels within 1 to 2 hours; they are eliminated rapidly. Time-release preparations have longer half-lives; a single dose may not reach peak levels for 4 to 8 hours after administration. Side-effects and toxicity are directly related to serum theophylline levels. The first symptoms of nausea, restlessness, headache, and vomiting may occur with serum levels as low as 13 g/ml. More serious toxic effects such as seizures or coma usually occur with serum concentrations greater than 40 μg/ml. Convulsions may occur in the absence of other symptoms of toxicity. Serum levels should be monitored, therefore, if there are signs of toxicity or failure to respond appropriately to the medications.

Drugs such as erythromycin and troleandomycin affect the metabolism of theophylline and may increase or decrease serum levels accordingly. Similarly, cigarette smoking and the presence of heart problems or liver dysfunction may affect theophylline metabolism.

Cromolyn sodium (Intal), introduced in the early 1970s, has become a valuable adjunct in the treatment of chronic asthma. A recent report (3) has listed its mode of action as the following: (1) prevention of mediator release from mast cells; (2) modulation of some types of reflex-induced bronchoconstriction; and (3) amelioration of nonspecific bronchial hyperreactivity in some asthmatic patients. Due to poor gastrointestinal absorption, cromolyn is administered by inhalation; patients may suffer only occasional side-effects such as coughing or rare urticarial and anaphylactic reactions (7). Because cromolyn is purely a prophylactic drug, it should never be administered for an acute attack of asthma, and should be discontinued during an exacerbation of wheezing.

Finally, patients whose asthma cannot be controlled with a combination of medications and/or environmental control measures may require treatment with corticosteroids, which should be administered under the close supervision of a physician. For a severe attack, intravenous Solu-Cortef or Solu-Medrol may be given every 4 to six hours at a dose of 1 mg/kg with a maximum of 40 mg per dose. For a less severe attack, prednisone may be given orally, up to 60 mg/day. This should be tapered off as quickly as possible. If maintenance therapy is necessary the lowest possible dose may be given on alternate days for as short a time as possible. This greatly reduces the risk of severe side-effects, such as adrenal suppression, osteoporosis, Cushing habitus, hypertension, peptic ulcer, cataracts, pseudotumor, or susceptibility to opportunistic microbial agents.

The mechanism of action of steroids in asthma, though not entirely understood, depends mostly on their anti-inflammatory activity and on their ability to potentiate the effect of adrenergic drugs on adenyl cyclase. Although steroids are known to stabilize cellular and lysosomal membranes, suppress synthesis of prostaglandins, and reduce stores of histamine and SRS-A (40), they do not affect immunologic release of mediators or inhibit their biological effects.

Beclomethasone dipropionate (Vanceril or Beclovent), a corticosteroid aerosol, became commercially available in the Unites States in 1976. It has proven to be beneficial in controlling asthma without producing the systemic adrenocortical suppression that results from oral steroid use (2). In most steroid-dependent asthmatics, beclomethasone use has permitted a significant decrease or complete discontinuation of their dependence on oral prednisone.

IMMUNOTHERAPY

It is unfortunate to note that in all reports published to date regarding the role of immunotherapy in the management of allergy patients, the authors studied only ragweed, grass, dust, mite, or cat antigens (15–17, 22,29,31). In 1980 Norman (28) reviewed the available literature and concluded that, for those antigens that have been studied with controlled clinical trials, immunotherapy is a valuable treatment modality. Because of the difficulty in performing these clinical trials, however, similar studies with other pollen and mould antigens may never be performed. Therefore, it is important that we understand the immunologic and physiologic mechanisms and changes induced by immunotherapy. We may then be able to deduce the efficacy of immunotherapy for treatment of mould allergy. Before this can be accomplished, we need accurate identification of the most important allergenic fungi, and the responsible antigens for each fungus.

Historically, immunotherapy for pollens was introduced by Noon in 1911 (27); immunotherapy for moulds, however, was not used until the 1930s (11), although a few European physicians, notably Van Leeuwen in Holland and Hansen in Germany, were interested in moulds as a cause of allergy before 1930. Bernton et al., in Washington, DC (4) were the first in the United States to test patients with various mould spores. Among several hundred patients tested, four patients with marked sensitivity to *Cladosporium* were found. By using an extract of *Cladosporium* for "desensitization," Bernton and colleagues achieved satisfactory alleviation of symptoms in three of four patients.

In 1935 Feinberg (11), in an attempt to analyze the prevalence of mould allergy and to identify those fungi important in seasonal and perennial allergic diseases, reported on 243 patients, 123 with seasonal asthma or hay fever and 120 with perennial symptoms. He tested all of them with extracts of *Alternaria, Aspergillus, Chaetomium, Penicillium, Mucor, Trichophyton, Candida,* and a few other fungi. In patients suffering seasonal allergies, Feinberg found 39 reactive patients, with most reacting to *Alternaria* extract. In patients suffering from perennial problems, only 21 patients were reactive to fungal extracts, with no specific fungus being dominant.

The inadequacies of mould extracts were soon recognized, and Browning reported in 1942 that most extracts used were irritants of no clinical value (8). As a result of this, Prince and associates organized the Association of Allergists for Mycological Investigation to develop mould extracts and to study their efficacy (33–35). Although since then many studies have been done to improve the quality of mould extracts, a recent report from Aas et al. comparing skin tests and radioallergoabsorbent tests (RAST) of five mould extracts from different manufacturers showed considerable differences in allergenic potency and composition (1).

Methods are currently being developed to standardize allergenic extracts of moulds using RAST testing, crossed immunoelectrophoresis (CIEP), crossed radioimmunoelectrophoresis (CRIE), isoelectric focusing, total hemolytic complement activation, and leukocyte histamine release (13,21,42). The objectives of this standardization are to identify the major and minor antigens for each extract, purify these antigens, and ensure the reproducibility of allergenic activity from one batch to another. It is also important to remove irritants from these antigens and to identify time and storage limitations on perishability.

In 1976 Yunginger et al. developed methods to study the allergenic components of *Alternaria* extract prepared from the mycelial growth of the fungus (47,48). They reported in 1980 that the major allergenic fraction is a glycoprotein, which they called Alt-1 (46). Hoffman et al. have discovered that there are at least eight spore-specific antigens not present in the mycelial extracts of *Alternaria* (14). They found that half of the *Alternaria*-sensitive patients were significantly more reactive to these spore-specific antigens than to the commercial mycelial extract, indicating that the latter extract is inadequate. Similar attempts were also made by Karr et al. to identify the antigen components of various species of *Aspergillus* (18).

It is easy to understand the difficulty in evaluating the success of immunotherapy for mould allergy. Not only is the potency of the extracts unreliable, but current extraction procedures fail to include spore-specific fractions that may be important for some mould sensitive patients.

Another problem is the difficulty for the average allergist to identify clinically relevant fungi. Most allergenic fungi belong to the group Deuteromycetes. It has been demonstrated, however, that smut and rust spores (Basidiomycetes) may also cause allergic reactions (23,36) even though they are rarely used for skin testing. Although the Basidiomycetes fungi are abundant when the humidity is high in summer and fall, their actual relationships to asthma attacks have yet to be explored.

The procedure of immunotherapy was developed after many years of clinical research. Even though general practice standards have been outlined by Patterson (31), individual allergists may utilize minor variations. In summary, after a comprehensive history has been taken and a physical examination done, appropriate skin testing will help in identification of the aeroallergens responsible for the patient's problem, whether allergic rhinitis or IgE-mediated asthma. The immunotherapy should be initiated with a dose of the appropriate aqueous mould extract 10 to 100-fold lower than the concentration needed to elicit a positive intradermal skin test. The subcutaneous injections should be administered weekly at increasing volumes and concentrations to a maximum tolerated dose. This maintenance dose, which is usually in the range of 2500 to 5000 protein nitrogen units (PNU), should be administered at lengthening intervals until it is given monthly, as long as it achieves amelioration of the patient's symptoms. The immunotherapy is usually continued for 3 to 4 years unless the patient has had no relief. In that case the responsible allergens, dosage of extracts, presence of immediate reaction, and other environmental causes of allergic disease should be evaluated carefully. Cessation or alteration of treatment in regard to dosage and composition of the extract should then be determined.

Recently a form of rush immunotherapy, based on the schedules used in hymenoptera venom desensitization, was evaluated for perennial childhood asthma by Goldstein and Chai (12). Their protocol consisted of four dilutions of extract: 10, 100, 1000, and 5000 PNU/ml, with up to five antigens mixed in the same vial. The volume increased from 0.05 to 0.5 ml for each dilution and the injections were given twice daily, 5 days per week, for the lowest two dilutions; injections were then given once per week for the fourth dilution, until the maintenance dose of 0.5 ml of 5000 PNU was achieved. Thereafter, patients received weekly injections of that maintenance dose.

Nine patients completed this protocol with 14 antigens, including dust, mite, *Alternaria,* and *Hormodendrum.* Results of bronchoprovocation testing before and after immunotherapy showed marked improvement in bronchial reactivity. This study demonstrated the efficacy of rush immunotherapy, as well as the significant response of a few patients to immunotherapy with *Alternaria* and *Hormodendrum.*

There have been reports of immunotherapy with polymerized pollen extracts (32) and intranasal immunotherapy with ragweed antigens (26); no such trials, however, have been attempted with mould extracts.

ENVIRONMENTAL CONTROL

In the case of mould allergy, isolation of the patient from the cause is the single most efficacious form of therapy we can advise. This was recognized by Blackley in his 1873 treatise:

> I (have never) met with a case in which I could feel sure that the administration of remedies had really produced a cure. It is true many cases are given by authors where the use of certain remedies seemed to be followed by an improvement or by a cessation of the symptoms, but in most of these cases I am convinced that the cure was due to the patient's removal beyond the reach of the cause or to the gradual diminution in the quantity of the latter (5).

Because fungi are so prevalent in our environment, isolation of the patient is a difficult task, but certain measures can and must be taken to lessen a patient's environmental exposure to moulds, whether indoors or outdoors.

Winter snow, rainfall, and cold weather almost eliminate fungal spores in the atmosphere, but the rising temperatures of spring produce conditions favorable for mould proliferation. Fungi grow especially well on decaying plant matter when humidity is high. Therefore, mould-sensitive people should avoid dense foliage and decaying vegetation found in wooded areas in spring and fall. They should not cut grass or rake leaves without the protection of spore-filtering masks.

Kozak et al. (19,20) have demonstrated that conditions of high shade, dense overgrown landscaping, and organic debris around homes and buildings significantly increase the indoor mould counts. Therefore, foundation shrubs should be pruned regularly and accumulated leaves should be removed. Ivy growing on walls or overgrown ground cover should be eliminated.

Indoors there are three conditions that promote mould growth and increase the amount of fungal spores: darkness, moisture and accumulation of dust (39).

Dust can be controlled by removal of stuffed furniture, toys, and carpeting, and by encasing mattresses in allergen-proof covers (Allergen-Proof Encasings). The filters of furnaces and air conditioners also trap and disperse spores and should be cleaned regularly. Treating filters with an active germicidal or fungicidal solution after cleaning can help control the number of fungal spores. Electrostatic air filters (made by Honeywell, Sears, and others) and Hepa filters (such as Vitaire Air Purifier, Allergen-Proof Encasings) effectively filter mould spores as well as dust and pollens from the indoor air.

Damp and dark areas, such as basements and attics, can be found in any home, school, or private or public building. These areas need careful

Fig. 11–1. Mould growth secondary to moisture seeping through basement floors and walls.

Fig. 11–2. Mould colonies growing on rungs of chair stored in basement area over a summer.

and frequent cleaning and dehumidification to eliminate any fungal growth (Fig. 11–1). Kitchens and bathrooms, which are frequently moist and warm, must be well aerated and may be cleaned with an antifungal agent such as bleach (Clorox) or hyperchlorite (X-14). Furniture, boxes, books, and other paraphernalia stored in damp areas provide multiple surfaces for mould proliferation (Fig. 11–2).

Basements and crawl spaces are naturally damp and dark, providing a perfect environment for mould growth (Fig. 11–3). Basement walls and floors should be painted with a moisture-proof sealer. Proper drainage of moisture away from the foundation of the house must be ensured by keeping gutters and downspouts functional and by proper grading of the land around the house. Crawl spaces should be filled with a 3-inch layer of stones and then covered with a plastic sheet. Dehumidification with an electric dehumidifier should be used continuously in damp areas. Attics can be protected from moisture by repairing all roof leaks and by using vapor barrier insulation.

Certain household appliances can cause problems, either from accumulated moisture or as a source of growing and disseminating fungi.

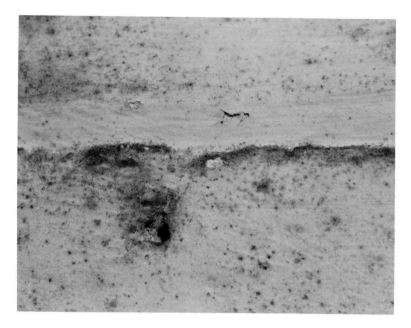

Fig. 11-3.   Fungal colonies growing on painted brick wall in basement.

Fig. 11-4.   A fungal colony growing on a refrigerated pepper.

Clothes washers and improperly vented dryers can greatly increase indoor humidity. Refrigerators and freezers with leaky door gaskets can build up moisture on their outer surfaces on which moulds will grow. In the refrigerator, fruits and vegetables left for long periods are a good substrate for fungal growth (Fig. 11–4). Growth of fungi in or around refrigerators can be controlled partially by rubbing refrigerator gaskets with vegetable oil to help retain a proper seal. Cool-mist humidifiers used to ease the breathing of allergic patients and evaporative coolers can actually compound allergy problems by aerosolizing fungal spores that grow in the water and on the pads. Solomon has reported 100-fold increases in the number of fungal particles recovered in bedroom air samples during operation of cool-mist vaporizers contaminated with mould (38).

The home contains many areas in which mould spores can proliferate. The difficulty in proper home maintenance is evident, but the relief to the patient is so great that constant diligence must be maintained.

REFERENCES

1. Aas, K., et al.: Immediate-type hypersensitivity to common moulds. Allergy, **35**:443, 1980.
2. Bacal, E. and Patterson, R.: Long-term effects of beclomethasone dipropionate on prednisone dosage in the corticosteroid-dependent asthmatic. J. Allergy Clin. Immunol., **62**:72, 1978.
3. Bernstein, I.L.: Cromolyn sodium in the treatment of asthma: Changing concepts. J. Allergy Clin. Immunol., **68**:247, 1981.
4. Bernton, H.S., et al.: The role of *Cladosporium*, a common mold in allergy. J. Allergy, **8**:363, 1937.
5. Blackley, C.: Experimental Research on the Cause and Nature of Catarrhus Aestivus. London, Bailliere, Tindall and Cox, 1873. (Revised edition, London, Dawson Pall Mall, 1959.)
6. Brown, H.M., Storey, G., and Jackson, F.A.: Beclomethasone dipropionate aerosol in treatment of perennial and seasonal rhinitis. A review of five years' experience. Br. J. Clin. Pharmacol., **4**:283, 1977.
7. Brown, L.A., et al.: Immunoglobulin E-mediated anaphylaxis with inhaled cromolyn sodium. J. Allergy Clin. Immunol., **68**:416, 1981.
8. Browning, W.H.: Mold fungi in the etiology of respiratory allergic diseases. II. Mold extracts, a statistical study. J. Allergy, **14**:231, 1943.
9. Cohan, R.H., et al.: Treatment of perennial allergic rhinitis with cromolyn sodium. Double-blind study on 34 adult patients. J. Allergy Clin. Immunol., **58**:121, 1976.
10. Clayton, D.E., et al.: Short-term efficacy trial and twenty-four-month follow-up of flunisolide nasal spray in the treatment of perennial rhinitis. J. Allergy Clin. Immunol., **67**:2, 1981.
11. Feinber, S.M.: Mold allergy: Its importance in asthma and hay fever. Wis. Med. J., **34**:254, 1935.
12. Goldstein, G., and Chai, H.: Efficacy of rush immunotherapy in decreasing bronchial

sensitivity to inhaled antigens in perennial childhood asthma. Ann. Allergy, **47**:333, 1981.
13. Grimmer, O.: A practical approach to allergen standardization. Allergy, **35**:220, 1980.
14. Hoffman, D., et al.: Isolation of spore-specific allergens from *Alternaria*. Ann. Allergy, **46**:310, 1981.
15. Johnstone, D.: Immunotherapy in children: Past, present, and future (Part I). Ann. Allergy, **46**:1, 1981.
16. Johnstone, D.: Immunotherapy in children: Past, present, and future (Part II). Ann. Allergy, **46**:59, 1981.
17. Johnstone, D., and Dutton, A.: The value of hyposensitization therapy for bronchial asthma in children—a 14-year study. Pediatrics, **42**:793, 1968.
18. Karr, R., et al.: An approach to fungal antigen relationships by radioallergosorbent test inhibition. J. Allergy Clin. Immunol., **67**:194, 1981.
19. Kozak, P., et al.: Factors of importance in determining the prevalence of indoor molds. Ann. Allergy, **43**:88, 1979.
20. Kozak, P., et al.: Currently available methods for home mold surveys. II. Examples of problem homes surveyed. Ann. Allergy, **45**:167, 1980.
21. Kwong, F.: Allergen extracts and purified allergen in immunotherapy. Ann. Allergy, **47**:162, 1981.
22. Lichtenstein, L.: Editorial: An evaluation of the role of immunotherapy in asthma. Am. Rev. Resp. Dis., **117**:191, 1978.
23. Lopez, M., Salvaggio, J., and Butcher, B.: Allergenicity and immunogenicity of Basidiomycetes. J. Allergy Clin. Immunol., **57**:480, 1976.
24. Mygind, N.: Mediators of nasal allergy. J. Allergy Clin. Immunol., **70**:149, 1982.
25. Neuman, I., and Toshner, D.: Beclomethasone dipropionate in pediatric perennial extrinsic rhinitis. Ann. Allergy, **40**:346, 1978.
26. Nichelsen, J., et al.: Local intranasal immunotherapy for ragweed allergic rhinitis. I. Clinical response. J. Allergy Clin. Immunol., **68**:33, 1981.
27. Noon, L.: Prophylactic inoculation for hay fever. Lancet **1**:1572, 1911.
28. Norman, P.: An overview of immunotherapy: Implications for the future. J. Allergy Clin. Immunol., **65**:87, 1980.
29. Norman, P., and Lichtenstein, L.: The clinical and immunologic specificity of immunotherapy. J. Allergy Clin. Immunol., **61**:370, 1978.
30. Norn, S., and Skov, P.S.: The pharmacological basis of drug treatment in bronchial asthma. Allergy, **35**:549, 1980.
31. Patterson, R.: Clinical efficacy of allergen immunotherapy. J. Allergy Clin. Immunol., **64**:155, 1979.
32. Patterson, R.: Allergen immunotherapy with modified allergens. J. Allergy Clin. Immunol., **68**:85, 1981.
33. Prince, H.E.: Mold fungi in etiology of respiratory allergic diseases: Immunological studies with mold extracts; skin tests with broth and washings from mold pellicles. Ann. Allergy, **2**:500, 1944.
34. Prince, H.E., et al.: Mold fungi in etiology of respiratory allergic diseases: Further studies with mold extracts. Ann. Allergy, **7**:301, 597, 1949.
35. Prince, H.E., Tatge, E.G., and Morrow, M.B.: Mold fungi in etiology of respiratory allergic diseases: Further studies with mold extracts. Ann. Allergy, **5**:434, 1947.
36. Salvaggio, J., and Aukrust, L.: Mold-induced asthma. J. Allergy Clin. Immunol., **68**:327, 1981.
37. Secher, C., et al.: Significance of H_1 and H_2 receptors in the human nose: Rationale for topical use of combined antihistamine preparations. J. Allergy Clin. Immunol., **70**:211, 1982.
38. Solomon, W.R.: Fungus aerosols arising from cold-mist vaporizers. J. Allergy Clin. Immunol., **54**:222, 1974.
39. Solomon, W.R.: Assessing fungus prevalence in domestic interiors. J. Allergy Clin. Immunol., **56**:235, 1975.
40. Stein, M.: New Directions in Asthma. Park Ridge, IL., College of Chest Physicians, 1975.
41. Taylor, G., and Shivalkar, P.R.: Disodium cromoglycate: Laboratory studies and clinical trials in allergic rhinitis. Clin. Allergy, **1**:189, 1971.
42. Weeke, B.: Standardization of allergen preparations. Allergy, **35**:172, 1980.

43. Weinberger, M.: Theophylline for treatment of asthma. J. Pediatr., **92**:1, 1978.
44. White, B.H., and Daeschner, C.W.: Aminophylline (theophylline ethylenediamine) poisoning in children. J. Pediatr. **49**:262, 1956.
45. Wyatt, R., Weinberger, M., and Hendeles, L.: Oral theophylline dosage for the management of chronic asthma. J. Pediatr., **92**:125, 1978.
46. Yunginger, J.W., et al.: Studies on *Alternaria* allergens. III. isolation of a major allergenic fraction (Alt-1). J. Allergy Clin. Immunol., **66**:138, 1980.
47. Yunginger, J.W., Jones, R.T., and Gleich, G.J.: Studies on *Alternaria* allergens. II. Measurement of the relative potency of commercial *Alternaria* extracts by the direct RAST and by RAST inhibition. J. Allergy Clin. Immunol., **58**:405, 1976.
48. Yunginger, J.W., Roberts, G.D., and Gleich, G.J.: Studies on *Alternaria* allergens. I. Establishment of the radioallergosorbent test for measurement of *Alternaria* allergens. J. Allergy Clin. Immunol., **57**:293, 1976.

Chapter 12

BRONCHOPULMONARY ASPERGILLOSIS AND ASPERGILLOMA

Gerald E. Wagner

The Latin derivative of the genus name *Aspergillus* comes from *asperge,* meaning "to scatter." The name is certainly justified, because this fungus is probably one of the most common microorganisms found in any environment. There are more than 150 species and varieties in the genus *Aspergillus,* and their ubiquity is due largely to the characteristics of their asexual sporulation structure. The conidia, or spores, are arranged on flask-shaped structures called sterigma that are, in turn, located on the vesicle. The number and arrangement of the sterigma on the vesicle and of the conidia on the sterigma are employed in differentiating the various species of this genus. The vesicle is at the end of a specialized hypha, the conidiophore; it usually extends above the growth substrate that supports the mycelial mass. The conidia are easily dislodged from the sporulation structure and are dispersed by wind currents, animals, and fomite.

In addition, the aspergilli have the ecologic advantage of being able to use a great variety of simple and complex substrates as nutrients for growth. The apergilli have been isolated from "the winds of the Sahara to the snows of Antarctica." The fungus has been found growing in industrial sulfuric acid-copper sulfate plating baths and on refrigerator walls. This tenacious microorganism grows in formalinized tissue specimens as well as in plain tap water that has been left standing for a few days.

As might be expected, the ubiquity of the aspergilli leads to almost constant human exposure to the organism. The most frequent mechanism of entry of the aspergilli into the human body is through inhalation of spores, although contamination of wounds and ingestion of spores may also result in infection. Of the large number of species of aspergilli, only about 12 species have been routinely reported as the etiologic agents of human infections; the most common species are *Aspergillus fumigatus*, *A. flavus*, and *A. niger*. In addition, *A. fumigatus*, among others, causes a substantial loss of both domestic and wild animal life, especially of water fowl. Wild ducks and other water fowl appear to develop pulmonary aspergillosis as the result of their underwater scavenging. The aspergilli are also important agents in the spoilage of foodstuffs and in the decay and destruction of other materials.

In reference to medical microbiology, recognition of the aspergilli as causative agents of human disease is relatively ancient. In 1847 Slutyer accurately described human aspergillosis (49). Less than a decade later, the renowned German pathologist Virchow so distinctly characterized the etiologic agent of four cases of bronchial and pulmonary aspergillosis as to allow contemporary identification of the fungus as *A. fumigatus* (56).

In the broadest and most important sense, there are two categories of human aspergillosis: a noninvasive form, often characterized by an allergic response, and an invasive form, characterized by the growth of hyphae into host tissue. The pathologic processes of aspergillosis may be delineated further under these two headings into five categories:

1. Toxic reactions due to the ingestion of food contaminated with the fungus or metabolites of the fungal growth; an example is the production of aflatoxin, a carcinogenic mycotoxin.
2. Allergy or allergic sequelae caused by the inhalation of spores or the transient growth of the fungus in body orifices.
3. Colonization of preformed cavities or necrotic tissue without active extension of the hyphae into healthy tissue.
4. Invasive granulomatous disease of lungs or other organs, which results in inflammation and necrosis.
5. Systemic, usually fatal, disseminated disease.

Within the scope of this chapter, the second and third categories will be discussed in terms of bronchopulmonary aspergillosis and aspergilloma.

BRONCHOPULMONARY ASPERGILLOSIS

The inhalation of spores of the aspergilli may result in intermittent obstruction of the bronchi in atopic individuals. This condition has classically been referred to as extrinsic asthma, although this term is no longer considered completely adequate. Allergic bronchopulmonary aspergillosis, a type of extrinsic asthma, was first described by Hinson and associates in 1952 (22). The criteria used for the diagnosis of allergic bronchopulmonary aspergillosis, however, still vary to some degree (15). In the United Kingdom this diagnosis is made if an asthmatic individual is found to have transient pulmonary infiltrates, eosinophilia of the peripheral blood, and a positive immediate hypersensitivity (type I) reaction to an extract of *Aspergillus fumigatus* applied by a scratch skin test (33). The diagnostic criteria within the United States, in addition to those just mentioned, include the demonstration of precipitating antibodies against aspergillus antigens and an increased serum concentration of IgE (43).

Additional criteria may include the presence of classic central bronchiectasis, culture of *Aspergillus* from sputum, the expectoration of sputum plugs, and the development of a late (type III) skin test reaction to aspergillus antigens. A range of 10 to 20% positive reactions has been detected in routine skin testing of asthmatic patients in whom a commercially prepared carbol-saline extract of *A. fumigatus* mycelia was used (20). A low but significant percentage of asthmatic patients gave positive precipitating antibody tests when crude extracts of aspergillus antigens were employed (30).

Most individuals with asthma due to the inhalation of spores of the aspergilli are typical atopic subjects suffering from bronchial symptoms as the result of various allergic sensitivities. There is also a group of patients in this category who demonstrate a seasonal aggravation of the asthmatic condition, between October and February (39). Recurrent infiltrations appear at any site within the lungs, although the most common site is near the medium-sized bronchi. The repeated asthmatic "attacks" generally increase in severity of symptomatology, and the infiltrates may last days or even weeks. Ellis (16) has reported that the infiltrates may be peribronchial or may involve collapse of a portion of a lobe, an entire lobe, or even the whole lung.

The inhaled spores may germinate in the bronchi and sputum plugs, but the hyphae do not invade the surrounding tissue. A great amount of antigens is probably released during the formation of hyphae, resulting in a localized allergic response in the bronchi and bronchial walls. This is supported by the fact that the bronchi demonstrate normal filling peripheral to characteristic bronchiectasis (46). Fibrosis of the upper lobe of the lung is not uncommon in patients with long-standing recurrent allergic bronchopulmonary aspergillosis (20). Therefore, it has been suggested that any individual with a history of recurrent pneumonic episodes should be evaluated for allergic bronchopulmonary aspergillosis.

It is believed that approximately 70% of early onset asthmatics exhibit a type I hypersensitivity reaction to exogenous antigens such as spores of the aspergilli or to other so-called "organic dusts." In contrast, only about 30% of the late onset asthmatic patients react to extrinsic allergens. In reality it may be difficult to determine whether or not the late onset asthmatic belongs to a low atopic status group with isolated sensitivity to aspergillus antigens, or to a nonatopic group with precipitating antibodies to an isolated antigen. Information on IgE levels may be helpful in making this differentiation. In a British study (39) using prick skin test reactions and inhalation tests, only a very small number of late onset asthmatics responded with immediate (type I) hypersensitivity reactions to inhalation tests and only a few had a strong skin test reaction, indicating that they were atopic individuals with allergic bronchopulmonary aspergillosis. These few individuals also lacked precipitating antibodies.

Bardana and coworkers have reported using a number of test systems for detecting antibodies against aspergillin prepared from sonically disrupted hyphae of *Aspergillus fumigatus* in the sera of 28 patients with allergic bronchopulmonary aspergillosis. The radioimmune assay was employed as a quantitative method for measuring aspergillus antibody. The test was based on concentration of circulating antibody; it was used to differentiate antibody stimulation by continual environmental exposure from that of chronic infection by *Aspergillus* spp. Cross reactivity with various antigenic determinants in related and unrelated microorganisms presented a problem; the investigators felt, however, that this test system was reliable in detecting patients with allergic bronchopulmonary aspergillosis when purified aspergillin was employed as the antigen. The titer of complement-fixing antibody to crude aspergillin was found to be much lower in patients with allergic bronchopulmonary aspergillosis than it was in those individuals with invasive disease. As expected, this difference is the result of lower amounts of IgG antibody against aspergillus antigens in the atopic asthmatic individual.

The immunodiffusion test has been the most commonly used technique for detecting antibody against the aspergilli (5,6,8,10,20,26, 30,34,35,37,39,52,58). In the 28 patients with allergic bronchopulmonary aspergillosis studied by Bardana et al. (4), precipitin bands (usually single) were weak, if present at all, in concentrated serum. These investigators concluded that, although differential serologic diagnosis of allergic bronchopulmonary aspergillosis is difficult, the use of multiple tests (i.e., radioimmune assay, complement fixation test, and the immunodiffusion test) for the detection of antibody against *Aspergillus* spp. increases the chance of making the correct diagnosis.

Classic allergic bronchopulmonary aspergillosis is believed by some to be, in its simplest form, extrinsic bronchial asthma. The inhalation of spores of the aspergilli, particularly of *Aspergillus fumigatus*, results in entrapment of these spores in the bronchi, where they may germinate and form a relatively luxuriant growth of hyphae that could block the bronchial lumen; sporulation structures are rarely formed *in vivo* (Fig. 12–1). No host tissue invasion occurs, but the release and absorption of fungal antigens results in a reaginic IgE antibody response responsible for immediate (Type I) hypersensitivity reactions. This type of reactivity,

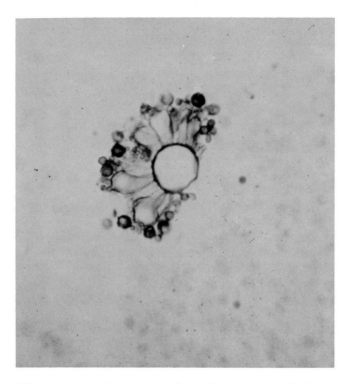

Fig. 12–1. Photomicrograph clearly shows the vesicle, sterigma, and a few spores of *Aspergillus* sp. in a pathologic preparation of lung tissue. (Courtesy of Dr. Morris A. Gordon.)

in turn, is believed to account for bronchospasms, eosinophilia, and immediate-type skin reactions in the patient with allergic bronchopulmonary aspergillosis. The individual may produce precipitating IgG and IgM antibodies, which are believed to be responsible for immune complex disease (type III). This reaction would account for the pulmonary infiltrates, damage to the bronchial walls, and late skin reactivity. Recent studies have suggested that cell-mediated hypersensitivity (type IV), antibody-dependent cellular cytotoxicity, and alternate pathways of complement activation may also be active in the immunopathogenesis of allergic bronchopulmonary aspergillosis (43).

Corticosteroids have proven to be effective therapeutic agents in allergic bronchopulmonary aspergillosis. The efficacy of corticosteroids appears to be primarily through alleviation of asthma and diminution of sputum production. Thus, the bronchi become a less suitable environment for the prolific growth of *Aspergillus* (44), and the fungus is more easily eliminated from the body (34). Corticosteroids may also help alleviate symptoms of the disease by inhibiting the inflammatory response of the host. Radiographic evidence attests to the effectiveness of these therapeutic agents in allergic bronchopulmonary aspergillosis

(20,26,34). The length of time that corticosteroids should be administered is not clear; recommendations have ranged from 6 months to full lifetime therapy.

ASPERGILLOMA

Species of the genus *Aspergillus*, particularly *A. fumigatus*, colonize damaged lung tissue such as that formed in open healed lesions of pulmonary tuberculosis. This noninvasive colonization is referred to as aspergilloma, mycetoma, or fungus ball. The aspergilli may also form fungus balls secondary to many other diseases, including sarcoidosis (17), bronchial cyst (11), malignant disease (9), asbestosis (22), histoplasmosis (41), lung abscess (51), bronchiectasis (7), lung fibrosis (12), pneumonia (1), and ankylosing spondylitis (28). Various occupational diseases, such as farmer's lung, malt workers's lung, and wig cleaner's lung, may in some instances be a pulmonary fungus ball caused by *Aspergillus* spp.

Aspergilloma, as suggested by Segratain (48) may be a primary type (bronchiectatic) aspergilloma in which *Aspergillus* is the only pathogen, or it may be a secondary-type aspergilloma in which the fungus is a secondary invader of a preformed cavity in the lung. Others (36,56), however, have doubted the existence of the primary aspergilloma, maintaining that there is always some pre-existing lung cavity or cyst in which the aspergilli invade the tissues and grow as saprophytes.

The aspergilloma basically consists of a relatively spherical mass of intertwined hyphae, amorphous debris, fibrin, and an occasional inflammatory cell. A "fungus ball" may be produced in the laboratory by growing *Aspergillus fumigatus* in shaking broth culture. These *in vitro* fungus balls are reportedly analagous in structure to the aspergillomas that occur in the lung (57,58).

Several reports have clarified the natural course of untreated aspergilloma (3,40,55). Within the aspergilloma a continuous cycle occurs, with growing fungal cells predominating at some times while dead fungal cells are more prevalent at other times. In part, the microenvironment within the cavity regulates this cycle. Dead fungal cells undergo regressive structural changes that lead to either softening and fragmentation or to focal calcification. The calcification process may remain focal or may eventually involve most of the hyphal mass. Accordingly, the "life cycle" of the aspergilloma has been divided into five stages (3).

Initial Stage. This stage is difficult to recognize because there are no clinical symptoms or radiographic signs. Diagnosis, therefore, is made on the histopathologic finding of a small mass or film of hyphae in a preformed pulmonary cavity (Fig. 12–2). In this stage the fungus is always viable.

Fully Developed Stage. The cavity containing the aspergilloma de-

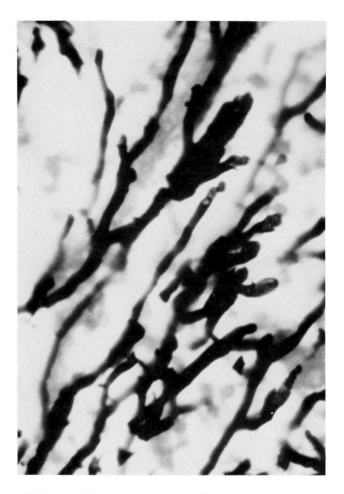

Fig. 12–2. Initial stage of formation of an aspergilloma shown in histopathologic section reveals a small mass (film) of hyphae. (Courtesy of Dr. Yousef Al-Doory.)

velops a thick fibrous wall during this stage (Fig. 12–3). It is also during this stage that the fungus goes through the cycle of cell growth and death. The different staining characteristics and loss of texture, including softening, fragmentation, liquefaction, and calcification, allow for the distinction between living and dead hyphae.

Calcified Stage. During this stage the entire fungus ball becomes calcified. This is believed to be an infrequent occurrence, because the dead hyphae usually liquefy and are expectorated with the sputum. When the fungus ball does remain in the cavity and becomes calcified, hyphae are recognizable in the stained decalcified aspergilloma.

Residual Stage. This stage may exist in two forms. One type is recognized by the presence of abscesses containing fragments of the fungus

Fig. 12–3. Photomicrograph shows a large mass of hyphae forming the border of an aspergilloma. (Courtesy of Dr. Yousef Al-Doory.)

ball that are revealed only by a careful search for hyphae in the cavity. The other form is characterized by the presence of tiny concretions in the bronchi (broncholiths) or in the pulmonary cavities (cavernoliths). These "small aspergillomas" may be free, or they may be incorporated into the granulation tissue of the cavity wall. Masses of septate hyphae characteristic of *Aspergillus* spp. are visible after decalcification and staining.

Abortive Stage. This stage is considered pathologically similar to the residual stage. The difference between the two stages is that the abortive stage is not preceded by the fully developed stage. This stage may represent the unsuccessful attempt of the fungus to grow and develop within the cavity.

The clinical picture of aspergilloma is considerably varied. Occasionally aspergilloma may be asymptomatic and resolve spontaneously (14,24). More commonly, however, the disease is characterized by hemoptysis. Levin (29) has estimated that hemoptysis occurs in half the patients with aspergilloma. More recently, though, other investigators have reported the incidence of this symptom to be approximately 75%. The hemoptysis can also be massive and fatal (13,23,27). The actual cause of the hemoptysis in aspergilloma remains unclear. It has been suggested that hemoptysis is due to the mechanical movement of the aspergilloma within the cavity, which causes trauma to the highly vascularized granulation tissue (40,55). Other causes of hemoptysis include release of an endotoxin of *Aspergillus fumigatus* that is responsible for the bleeding (22), and a type III hypersensitivity reaction that causes vascular damage leading to edema and hemorrhage (21).

A chronic cough, productive of sputum, is also a relatively common characteristic of aspergilloma. Occasionally generalized malaise and loss of weight may be symptomatic expressions of the disease. Less frequently, the systemic manifestation of fever, with no evidence of bacterial infection, may be present. Halvering et al. (21) have reported two patients with febrile illness and attributed it to a type III-mediated hypersensitivity reaction within the lung. Slowly progressing dyspnea may occur (12); except for the thickening of the pleura at the site of the cavity, however, the reason for dyspnea is not known. Secondary pneumonia due to either bacteria or to other opportunistic fungal pathogens also has been reported in patients with aspergilloma (25).

Aspergilloma nearly always occurs in the nonatopic individual. In some cases, however, asthma may develop at a later time and may be accompanied by pulmonary eosinophilia; a positive type I hypersensitivity reaction develops at this time. Additionally, aspergilloma may result from chronic allergic bronchopulmonary aspergillosis. Generally, the aspergilloma forms in an ectatic bronchus and the symptoms resemble those of allergic bronchopulmonary aspergillosis, except that severe bouts of hemoptysis are more common than in the latter.

Two well-documented cases of an allergic bronchopulmonary aspergillosis-like syndrome developing subsequent to, and as a consequence of, aspergilloma were reported by Ein et al. (15). There was no history of chronic asthma in either patient that could have resulted in pulmonary fibrosis and bronchiectasis, thereby creating a favorable environment for the development of a secondary aspergilloma. Both patients, however, did have a history of pulmonary tuberculosis that resulted in preformed cavities in which the aspergilli could colonize and develop into a fungus ball. The disease that subsequently developed was immunologically indistinguishable from allergic bronchopulmonary aspergillosis, with clinical and laboratory evidence of types I and III hypersensitivity reactions. Increased IgE levels were detected in the sera of the patients and, although it was specifically directed against *Aspergillus fumigatus*, the sera were also found to react with eight other antigens from other "organic dusts." One of the two patients showed significant improvement in pulmonary function when given corticosteroid (prednisone) therapy.

Avila (3) has performed immunologic studies on three patients diagnosed as having aspergilloma. Ouchterlony's gel diffusion technique, a microtechnique of double diffusion in agar, and immunoelectrophoresis were employed serologically to diagnose aspergilloma serologically. In two of the patients, each with calcified aspergilloma, precipitating antibodies against *Aspergillus fumigatus* were detected. The author explained that this was the result of the presence of live hyphae in the calcified aspergilloma producing an antigenic stimulus. In the third patient, diagnosed as having residual aspergilloma, no precipitating antibodies against *A. fumigatus* were detected. This result was explained by a lack of live fungus in the aspergilloma. Avila suggested that this result was analogous to that obtained in a patient who had an aspergilloma surgically removed. It was recommended that, in all cases of pulmonary aspergilloma lacking precipitating antibodies against *A. fumigatus*, a careful pathologic study be performed. Such a study should reveal the species of *Aspergillus* responsible for the aspergilloma and show the clinical stage of the disease.

Pepys (39), in a monograph on allergy to fungi, stated that the test for precipitating antibodies is almost always positive in patients with radiographic evidence of an aspergilloma. The author reported the results of a survey in which 98% (56 of 57) of the cases of aspergilloma were positive for precipitating antibodies, and 77% of these patients had six or more arcs on double diffusion tests. Such strong reactions have been considered to be of high diagnostic value in pulmonary aspergilloma (3,39,54).

Results of study by the British Tuberculosis Association (6) of 544 patients with open-healed tuberculous cavities yielded a strong precipitating antibody reaction in 11% of patients; these individuals also showed radiographic findings consistent with aspergilloma (5). An additional 4% of patients with findings suggestive of aspergilloma also showed strong precipitating antibody reactions. In cases in which the aspergilloma is surgically removed, dies, or is expectorated, the precipitating antibody test rapidly becomes weak or even negative (39).

Longbottom et al. (31) have reported strong precipitating antibody reactions in aspergilloma caused by *Aspergillus flavus, A. niger,* and *A. nidulans.* Precipitating antibody directed against *A. fumigatus* antigens was either absent or weak in these sera. One patient with aspergilloma caused by *A. nidulans* also had a positive precipitin reaction for *A. fumigatus,* suggesting that mixed infection by different species of *Aspergillus* may occur in aspergilloma (19) or, alternatively, suggesting cross reactivity between these two species.

Bardana et al. (4) measured serum immunoglobulin levels in 41 patients with aspergilloma and compared them to the levels in 100 normal healthy individuals. The IgG and IgA levels were high in the aspergilloma patients, being 2015 ± 540 mg/dl and 420 ± 157 mg/dl, respectively, as compared to 1274 ± 280 mg/dl and 227 ± 53 mg/dl for the controls. IgM levels were a little low, 67 ± 51 mg/dl in the patients with aspergilloma as compared to the control value of 127 ± 46 mg/dl. IgE levels were in relatively good agreement in both groups, with values of 178 ± 176 mg/dl in patients with aspergilloma and 199 ± 211 mg/dl in control sera. The C3 component of complement was also measured in these patients; it was 153 ± 40 mg/dl in the cases of aspergilloma and 145 ± 30 mg/dl in controls. This series of values appears to be useful in differentiating aspergilloma from allergic bronchopulmonary aspergillosis.

Controversy surrounds the issue of best mode of treatment for aspergilloma. Therapy with the most widely used systemic antifungal compound, intravenously administered amphotericin B, has not proven effective (18). Surgical excision of all aspergillomas often has been recommended (2,3,8,26,50), while others have recommended surgery only for those patients with hemoptysis (24,42). A more conservative approach based on careful observation has also been advocated by Varkey and Rose (53), who analyzed 15 cases of aspergilloma over an 11-year period. The patients in the study were followed for an average of 50 months. Of the eight untreated patients, the aspergilloma spontaneously disappeared in four, remained unchanged in three, and enlarged in only one.

Other surveys (37,47) have reported spontaneous clearing of the aspergilloma in 7 and 10% of patients; elimination of the fungus ball generally occurred within 36 months of the initial diagnosis. The risk of aspergilloma progressing to invasive or hematogenously disseminated disease is considered negligible (45). In addition, the therapeutic use of corticosteroids for underlying conditions does not appear to increase the chances of dissemination of aspergilloma (24) as often happens in other mycotic diseases. In fact, corticosteroid therapy may actually result in improvement in these cases (12). The progression and outcome of pulmonary aspergilloma appear to be principally related to the severity and extent of underlying disease.

ADDITIONAL COMMENTS

The number of voluntarily reported deaths due to *Aspergillus* species in the United States each year is small, and probably truly represents the "tip of the mycotic iceberg." The number of work hours lost and the discomfort due to allergic aspergillosis is undeterminable. Considering the nature of allergic aspergillosis, therapy is basically either supportive or is aimed at inhibiting the host's immune response. Standard antifungal compounds such as amphotericin B, nystatin, and 5-fluorocytosine have limited efficacy against the aspergilli in general and would be of little value in the treatment of either allergic bronchopulmonary aspergillosis or aspergilloma. In lieu of several years of investigation on the mode of action of antifungal compounds in medically important *Aspergillus* spp., it is my contention that effective treatment of aspergillosis is the result of limiting the growth of the fungus until the host's immune system can eliminate the infection.

There has not been as great an advancement made in the serologic diagnosis and differentiation of aspergillosis as there has been with other mycotic diseases. A major difficulty, of course, is the ubiquity of the aspergilli and the probability that many individuals have developed antibodies against this fungus at some time without overt disease. Another problem in serodiagnosis is the antigen richness of the fungi, and of *Aspergillus* spp. in particular, that leads to the high degree of cross reactivity with other fungi and with bacteria. As techniques become more sophisticated and less cumbersome and costly, it would be reasonable to expect that the production and use of monocolonal antibodies against specific *Aspergillus* antigens may play an important role in the serologic differential diagnosis of aspergillosis.

REFERENCES

1. Abbot, J.W., et al.: Pulmonary aspergillosis following post-influenzal bronchopneumonia treated with antibiotics. Br. Med. J., 1:523, 1952.
2. Alsam, P.A., Eastridge, C.E., and Hughes, F.A.: Aspergillosis of the lung—an 18-year experience. Chest, 59:28, 1971.
3. Avila, R.: Immunological study of pulmonary aspergilloma. Thorax, 23:144, 1968.
4. Bardana, E.J., et al.: The general and specific humoral immune response to pulmonary aspergillosis. Am. Rev. Resp. Dis., 112:799, 1975.
5. British Tuberculosis Association Research Committe: Aspergillus infection of persistent lung cavities after tuberculosis. Tubercle, 49:1, 1968.
6. Brownestam, R., and Hallberg, T.: Precipitins against an antigen extract of *Aspergillus fumigatus* in patients with aspergillosis or other pulmonary diseases. Acta Med. Scand., 177:385, 1965.
7. Bruce, R.A.: A case of pulmonary aspergillosis. Tubercle, 38:203, 1957.
8. Campbell, M.J., and Clayton, Y.M.: Bronchopulmonary aspergillosis: A correlation of the clinical and laboratory findings in 272 patients investigated for bronchopulmonary aspergillosis. Am. Rev. Resp. Dis., 89:186, 1979.
9. Carbone, P.P., et al.: Secondary aspergillosis. Ann. Intern. Med., 60:556, 1964.
10. Coleman, R.M., and Kaufman, L.: Use of the immunodiffusion in the serodiagnosis of aspergillosis. Appl. Microbiol., 23:301, 1972.
11. Corpe, R.F., and Corpe, J.A.: Bronchogenic cystic disease complicated by unsuspected choleresis and aspergillus infestation. Am. Rev. Tuberc., 74:92, 1956.
12. Davies, D., and Sommer, A.R.: Pulmonary aspergillosis treated with corticosteroids. Thorax, 27:156, 1972.
13. Donaldson, M.J., Koerth, C.J., and McCorble, R.S.: Pulmonary aspergillosis. J. Lab. Clin. Med., 27:740, 1942.
14. Duroux, A.: Lyse spontanée d'un aspergillome. J. Radiol., 45:647, 1964.
15. Ein, M.E., Wallace, R.J., and Williams, T.W.: Allergic bronchopulmonary aspergillosis-like syndrome consequent to aspergilloma. Am. Rev. Resp. Dis., 119:811, 1979.
16. Ellis, R.H.: Total collapse of the lung in aspergillosis. Thorax, 20:118, 1965.

17. Gorske, K.G., and Fleming, R.J.: Mycetoma formation in cavitary pulmonary sarcoidosis. Radiol., **95**:279, 1970.
18. Hammerman, K.J., Sarosis, G.A., and Tosh, F.E.: Amphotericin B in the treatment of saprophytic forms of pulmonary aspergillosis. Am. Rev. Resp. Dis., **109**:57, 1974.
19. Helluy, J.R., et al.: À propos d'un cas d'aspergilloma bronchique à *Aspergillus nidulans.* Ann. Med., **2**:142, 1963.
20. Henderson, A.H.: Allergic aspergillosis: Review of 32 cases. Thorax, **23**:501, 1968.
21. Hilvering, C., Stevens, M.A., and Orie, N.G.M.: Fever in aspergillus mycetoma. Thorax, **25**:19, 1970.
22. Hinson, K.F.W., Moon, A.J., and Plummer, N.S.: Bronchopulmonary aspergillosis— a review and a report of eight new cases. Thorax, **7**:317, 1952.
23. Husebye, K.D.: Serial studies in a case of pulmonary aspergillosis. Dis. Chest, **51**:327, 1967.
24. Israel, H.L., and Ostrow., A.: Sarcoidosis and aspergilloma. Am. J. Med., **47**:243, 1969.
25. Joynson, D.H.: Pulmonary aspergilloma. Br. J. Clin. Pract., **31**:207, 1977.
26. Kilman, J.W., et al.: Surgery for pulmonary aspergillosis. J. Thorac. Cardiovasc. Surg., **57**:642, 1969.
27. Krakowka, P., Rowanska, E., and Halwag, H.: Infection of pleura by *Aspergillus fumigatus.* Thorax, **25**:245, 1970.
28. Krohn, J., and Halvarsen, J.H.: Aspergilloma of the lung in ankylosing spondylitis. Scand. J. Resp. Dis. 63(Suppl.):131, 1968.
29. Levin, E.G.: Pulmonary intracavitary fungus ball. Radiol., **66**:9, 1956.
30. Longbottom, J.L., and Pepys, J.: Pulmonary aspergillosis: Diagnostic and immunological significance of antigens and C-substance in *Aspergillus fumigatus.* J. Pathol. Bacteriol., **88**:141, 1964.
31. Longbottom, J.L., Pepys, J., and Clive, F.T.: Diagnostic precipitin test in aspergillus pulmonary mycetoma. Lancet, **1**:588, 1969.
32. Macartney, J.N.: Pulmonary aspergillosis. A review and description of three new cases. Thorax, **19**:287, 1969.
33. Malo, J.L., Hawkins, R., and Pepys, J.: Studies in chronic allergic bronchopulmonary aspergillosis. I. Clinical and physiological findings. Thorax, **32**:254, 1977.
34. McCarthy, D.S., and Pepys, J.: Allergic bronchopulmonary aspergillosis. Clin. Allergy, **1**:261, 415, 1971.
35. McCarthy, D.S., and Pepys, J.: Pulmonary aspergilloma: Clinical immunology. Clin. Allergy, **3**:57, 1973.
36. Orie, N.G.M., de Vries, G.A. and Kilstha, A.: Growth of *Aspergillus fumigatus* in the human being. Am. Rev. Resp. Dis., **82**:649, 1960.
37. Parker, J.D., et al.: Pulmonary aspergillosis in sanitoriums in the South Central United States. Am. Rev. Resp. Dis., **101**:551, 1970.
38. Pecora, D.V., and Toll, M.W.: Pulmonary resection for localized aspergillosis. N. Engl. J. Med., **263**:785, 1960.
39. Pepys, J.: Hypersensitivity diseases of the lungs due to fungi and organic dusts. Monog. Allergy, **4**:1969.
40. Pimental, J.C.: Pulmonary calcification in the tumour-like form of aspergillosis: Pulmonary aspergilloma. Am. Rev. Resp. Dis., **94**:208, 1966.
41. Procknow, J.J., and Loewen, D.F.: Pulmonary aspergillosis with cavitation secondary to histoplasmosis. Am. Rev. Resp. Dis., **82**:101, 1960.
42. Riley, E.A., and Tennebaum, J.: Pulmonary aspergilloma or intracavitary fungus ball. Report of five cases. Ann. Int. Med., **56**:896, 1962.
43. Rosenberg, M., et al.: Clinical and immunological criteria of the diagnosis of allergic bronchopulmonary aspergillosis. Ann. Int. Med., **86**:405, 1977.
44. Rosenberg, M., et al.: The assessment of immunological and clinical changes occurring during corticosteroid therapy for allergic bronchopulmonary aspergillosis. Am. J. Med., **64**:599, 1978.
45. Sarosi, G.A., et al.: Aspergillomas occurring in blastomycotic cavities. Am. Rev. Resp. Dis., **104**:581, 1971.
46. Scadding, J.G.: The bronchi in allergic aspergillosis. Scand. J. Resp. Dis., **48**:372, 1967.
47. Schwartz, J., Baum, G.L., and Straub, M.: Cavitary histoplasmosis complicated by fungus ball. Am. J. Med., **31**:692, 1961.

48. Segratain, G.: Infections by fungi that ordinarily are saprophytes. Lab. Invest., **2**:1046, 1962.
49. Slutyer, T.: De Vegetalibus organismi animalis parasitis. Diss. Inaug. Berlin, 1847, p.14.
50. Solit, R.W., et al.: The surgical implications of intracavitary mycetomas (fungus balls). J. Thorac. Cardiovasc. Surg., **62**:411, 1971.
51. Stevenson, J.G., and Reich, J.M.: Bronchopulmonary aspergillosis: Report of a case. Br. Med. J., **1**:985, 1957.
52. Suyuki, S., et al.: On the immunochemical and biochemical studies of fungi. Jap. J. Microbiol., **11**:269, 1967.
53. Varkey, B., and Rose, H.D.: Pulmonary aspergilloma. A rational approach to treatment. Am. J. Med., **61**:626, 1976.
54. Villar, T.G., Pimental, J.C., and Avila, R.: Some apects of pulmonary aspergilloma in Portugal. Dis. Chest, **51**:402, 1967.
55. Villar, T.G., Pimental, J.C., and Costa, M.F.E.: The tumour-like forms of aspergillosis (pulmonary aspergilloma). Thorax, **17**:22, 1962.
56. Wagner, G.F.: Metabolism of 5-fluorocytosine in medically important aspergilli. Ph.D. dissertation, Virginia Commonwealth Univ., 1977.
57. Young, R.C., et al.: Aspergillosis: The spectrum of the disease in 98 patients. Medicine, **49**:147, 1970.

Chapter 13

HYPERSENSITIVITY PNEUMONITIS

Viswanath P. Kurup

Joseph J. Barboriak

Jordan N. Fink

A recent increased interest in the health implications of work and home environments has resulted in identification of various diseases associated with noxious environmental agents. Hypersensitivity pneumonitis (HP) is one of the occupational lung diseases inadequately reported on or even unknown in the past that is becoming increasingly recognized as an important cause of pulmonary impairment.

Hypersensitivity pneumonitis, also known as extrinsic allergic alveolitis, is an immunologic lung disease caused by sensitization of susceptible individuals to various inhalant antigens. The disease process usually involves the middle and upper bronchial airways and the lung parenchyma, and represents an interstitial lymphocytic or granulomatous allergic reaction of the pulmonary tissue. Clinically, the disease is characterized by intermittent episodes of chills, fever, cough, and shortness of breath occurring 4 to 8 hours after inhalation of a specific sensitizing agent. Chest x rays reveal diffuse nodular infiltrates or interstitial fibrosis, depending on the stage of the disease process. Pulmonary function abnormalities range from severe diffusion defects to varying degrees of restriction and obstruction. The antigens involved are of microbial origin, organic dusts, or even simple chemical molecules, such as toluene diisocynate (TDI) or cromolyn sodium. Regardless of the etiologic agent, the similarity of lung pathology and symptomatology suggests a common pathogenesis.

HISTORICAL BACKGROUND

A disease in grain sifters and measurers resembling hypersensitivity pneumonitis was reported by Ramazzini in 1713 (76). He associated the disease with decay, weevils, and mould in the grain. Farmer's lung, the first classic and well-studied example of HP, was described in Great Britain in 1932 (12). Other cases of occupation-related HP were reported soon afterwards (17,30,72). Bagassosis, caused by inhaled mouldy bagasse containing thermophilic actinomycetes, was reported from various parts of the world (9,10). Mushroom worker's lung, caused by the inhalation of compost dust containing thermophilic actinomycetes, was described by investigators in the United States and the United Kingdom (8,83). HP due to thermophilic actinomycetes contaminating heating and humidification systems of offices and homes has been reported more recently (6,22,24,25,45,52,56). The study of antigenic components of mouldy hay was carried out by various workers in this country and abroad (17,40,73).

ANTIGENS INVOLVED

A number of inhaled organic dusts or other materials can sensitize and cause hypersensitivity pneumonitis in susceptible individuals. The various sources of etiologic agents and the popular names of the disease are given in Table 13–1. Several new antigens have been recognized in association with HP, but the symptoms and pathogenesis of the diseases remain the same.

Numerous species of fungi are implicated in HP. Other antigenic sources include protozoa, lower bacteria, chemicals, insects, and avian materials. The higher bacteria belonging to the group of thermophilic actinomycetes are the single major group of organisms causing HP. These bacteria resemble fungi in several respects: they are universally distributed and can utilize various substrates for their growth (52). Because of the important role of thermophilic organisms in HP, they will be discussed in detail. Other actinomycetes have also been reported as causing HP, but are less frequently observed (37).

PATHOPHYSIOLOGY

Both humoral and cellular immune responses are thought to be involved in the pathogenesis of hypersensitivity pneumonitis.

Humoral Immunologic Responses

The most characteristic immunologic feature of HP is the presence of serum precipitins against the offending antigen (72). In many cases the

TABLE 13–1. **Etiologic Agents of Hypersensitivity Pneumonitis**

Disease	Source or Environment	Antigen
Farmer's lung	Mouldy hay, mouldy corn	*Micropolyspora faeni, Thermoactinomyces candidus, T. vulgaris, Saccharomonospora viridis*
Bagassosis	Mouldy bagasse	*Thermoactinomyces sacchari, T. candidus, T. vulgaris*
Mushroom worker's lung	Mouldy vegetable compost	*Thermoactinomyces vulgaris, T. candidus, Saccharomonospora viridis, Micropolyspora faeni*
Ventilation systems pneumonitis	Contaminated forced air systems	*Thermoactinomyces candidus, T. vulgaris, Penicillium* spp.
Maple bark disease		*Cryptostroma corticale*
Malt worker's lung	Mouldy malt	*Aspergillus clavatus, A. fumigatus*
Sequoiosis	Mouldy redwood dust	*Graphium* spp., *Aureobasidium pullulans, Trichoderma* spp.
Cheese worker's lung	Cheese mould	*Penicillium caesi*
Suberosis	Mouldy cork	*Penicillium frequentans*
Sauna taker's lung	Mouldy tub	*Aureobasidium pullulans*
Lycoperdonosis	Puffballs	*Lycoperdon pyriformae, L. gemmatum*
Bird breeder's lung	Serum and excreta of pigeons, budgerigars (parakeets), and chickens	Avian proteins
HP due to chemicals	Drugs and environment contaminated with chemicals	TDI, cromolyn sodium, nitrofurantoin
Coffee worker's lung	Coffee beans	Coffee bean dust
Furrier's lung	Animal hair	Hair dust
Vineyard sprayer's lung	Spray fungicide	Copper-containing spray
Pituitary snuffer's lung	Pituitary snuff	Bovine and porcine dried posterior pituitary gland

immunologic reactivity is directed against more than one species of thermophilic actinomycete or fungi (99,101). The complex nature of the offending antigens usually limits precise quantification of the antibody response in patients. Generally the antibody concentrations are high enough to be detected by agar gel double diffusion (29,67). Immunoglobulin classes of antibodies belonging to IgG, IgM, and IgA against thermophilic actinomycetes have been detected in patient sera by radioimmunoassay methods (70). It should be noted, however, that approximately 50% of asymptomatic individuals exposed to the same organic dusts also show precipitating antibodies (29). The precipitating antibodies against the sensitizing organic dust in symptomatic patients and in exposed individuals without symptoms cannot be differentiated qualitatively or quantitatively, although results of recent studies using antigen-antibody crossed immunoelectrophoresis show some differences in the pattern of precipitin bands (95). Both asymptomatic and symptomatic individuals have been shown to have antibodies to the etiologic agents; on rare occasions, however, individuals with the disease but no antibodies have also been described (19). In some reports the failure of patients to show antibody reactions may be due to the use of insensitive methods or inappropriate test antigens.

Because of the presence of precipitins in both symptomatic and asymptomatic individuals, the role of the antibody in the pathogenesis of HP cannot be defined precisely. Immune complex diseases usually occur when the ratio of antibodies to antigens reaches the antigen excess zone. Hence, the absolute quantity of antibody is not as important as the inhalation of the proper quantity of antigen and the subsequent interaction with this antibody in the lung or in the blood. The immune complex formed in the blood is usually due to antibody excess, while the complex formed in the alveolus may be due to antigen excess and thus may be more important in the disease process.

The 4- to 8-hour delay in the appearance of symptoms following exposure to the offending antigens seen in HP is comparable to that observed in immune complex diseases. When HP patients are skin-tested with specific antigens, an immediate reaction may sometimes be produced. In addition, an Arthus-like reaction (4–8 hours) is usually observed (23,66,72). This late reaction begins with erythema and edema and may progress to central necrosis, but usually subsides within 24 hours. A true delayed response in the absence of an Arthus reaction has not been observed. Antigens, complement, and immunoglobulins were detected in lung biopsies (23,72). Peripheral leukocytosis is commonly observed during the acute phases of HP. Neutrophilic pulmonary infiltrates in the lung, however, have not been regularly observed in lung biopsies. This may be because most biopsies are not obtained in the early acute stage of the disease, and this phenomenon may therefore be overlooked.

Cell-Mediated Immunologic Responses

The role of cell-mediated immunologic response in patients with HP was not studied until recently, although granulomatous immunologic response in these patients has been known for some time. Sensitized T lymphocytes in the peripheral circulation of patients with HP have been recently reported by several investigators (1,11,27). These studies utilized migration inhibition factor (MIF) and antigen-induced blast transformation, both of which are indicators of cell-mediated immunity (61,62). *In vitro* testing of cell-mediated immunity was reported to be of use in discriminating symptomatic and asymptomatic individuals. MIF was produced by peripheral lymphocytes from a large proportion of patients exposed to specific antigens. Antigen-induced lymphocyte transformation is considered to be an *in vitro* test for activated T lymphocytes. In HP lymphocyte transformation was detected in the presence of specific antigens. In most cases there was a positive correlation between the *in vitro* cellular immune response and the presence of circulating precipitins. Results of various animal model studies also strongly support the theory that activated T cells are involved in the pathogenesis of HP (35,38,85).

Pathologic Findings

Pulmonary histologic findings of HP vary according to the stage at which the disease is evaluated. Earliest recognized changes include the presence of lymphocyte infiltration in the alveolar walls (34,78). The alveolar space may contain plasma cells, engorged alveolar macrophages, and foamy histiocytes. In some forms of HP, such as pigeon breeder's disease, the foam cells are more prominent and may occupy an interstitial as well as an intra-alveolar position. Sarcoidlike granulomas are frequently found within the infiltrates.

In some severe or chronic cases, fibrotic areas may be seen in the alveolar interstitium. Lymphocytes and plasma cells may be interspersed, but these are much less numerous than those that occur in acute forms of the disease. The walls of the bronchioles may be thickened by lymphocytic infiltration, and the lumen may be obstructed by granulation tissue (34,78). Eventually fibrosis becomes more evident, pulmonary architecture becomes distorted, and typical fibrosis and cyst formation develop. Widening of the intermembranous space and edema are also seen occasionally in various stages of the disease. As the disease progresses, advanced fibrosis and emphysema become characteristic features.

Host Factors Involved

The immunopathogenesis of HP is very complex and involves many factors. It is well known that all exposed individuals do not exhibit the pathologic response. Some show no response at all, while others demonstrate varying degrees of immunologic response. Hence, additional host factors that determine the individual susceptibility to the disease must be postulated.

Surprisingly enough, only 5 to 10% of individuals regularly exposed to the etiologic agents of HP contract the disease (15). Therefore, the inherent immunologic reactivity of the host is probably more important than exposure to the antigen(s). Individual differences in immune responsiveness would cause pathologic immune responses in some and lack of susceptibility in others. These differences may be genetically determined or may be acquired. Several reports have suggested no association between the HL-A antigens (histocompatibility antigens) and

TABLE 13–2. Classification of Immunologic Mechanisms

Type	General Term	Reactant	Mechanism of Tissue Damage
I	Anaphylactic	Reaginic antibody, primarily IgE	Mediators form IgE-sensitized mast cells
II	Cytotoxic	Ig or IgM antibody against cell surface antigens	Antibody with complement reacts on cell surface, destroys cells
III	Immune complex	Antigen-antibody complexes	Complexes fix complement, which attract leukocytes; mediators released by these cells produce inflammation
IV	Cell-mediated	T lymphocytes	Lymphocytes reacting with antigens release lymphokines, which produce cellular inflammation

(From Gell, P.G.H., and Coombs, R.R.A.: Clinical Aspects of Immunology. 2nd Ed. Oxford, Blackwell Scientific Publications, 1968.)

various forms of HP (5,28,81). Thus, it is possible that both genetic and environmental factors predispose individuals to the development of HP. The inflammatory reaction in the lung may act as an adjuvant and, together with the inherited immune response, may permit the production of activated T cells to the antigens. This explanation is preliminary, however, and requires additional confirmation.

Experimental Animal Models

Pulmonary response to HP has been classified under each type of immune injury described by Gel and Coombs (Table 13–2). Several investigators have reported that the lung is capable of sustaining all four types of immune injury and their combinations in response to inhaled antigens. Utilizing animal models, scientists have attempted to elucidate the pathologic mechanism and compare those with the wide spectrum of events in human HP. In animal models, both known etiologic agents and defined antigens have been employed. HP has traditionally been classified as type III, immune complex, or Arthus-like immunity in the lungs. In most animal models, however, the immune injuries were mixed between types I and III, or were predominantly types III and IV (33–36,64,75,79,80,82,100).

Types of Immunologic Responses

Natually occurring HP has been reported in cattle. Several investigators found that precipitins in cattle against mouldy hay or *Micropolyspora faeni* alone were only indicative of exposure to hay. Healthy cattle lose the precipitins while in the pasture and produce them again during the winter season when they are housed (75,100). Wilkie (100) concluded that type I immune injury leads to type III damage in his *M. faeni*-induced cattle model. Passive cutaneous anaphylaxis and hemagglutination were reported in the *M. faeni*-challenged cattle. MIF was also produced from peripheral leukocytes exposed to antigens, however, indicating a delayed reaction (100).

Using defined antigens to insufflate rabbits, Richerson (79,80) observed the histopathologic changes of the lung that were compatible with immediate hypersensitivity (type I), immune complex disease (type III), and cellular hypersensitivity (type IV). Hemorrhagic pneumonitis produced in rabbits was infrequently detected in human HP, except when lung biopsies were examined during the acute episode of the disease. We have observed in our laboratory rats after their exposure to pigeon dropping extract to have typical interstitial pneumonitis similar to that seen in humans with HP (82,86). In guinea pigs, these investigators in our laboratory produced acute hemorrhagic pneumonitis similar to that observed by Richerson using defined antigen. Using immunofluoresence, they found complement (C3) immunoglobulin and antigen bound to the lung tissue; they could also transfer the disease to normals by serum transfer. These findings clearly indicate an immune complex or type III reaction.

Experimental animal models for pigeon breeder's disease have been developed in rats, rabbits, and monkeys (34,35,64). Rats and monkeys developed chronic granulomatous inflammation of the lung while rabbits failed to develop lung inflammation. Monkeys showed immunoglobulin, C3, and antigen in lung biopsies; the disease could also be transferred to control normals by serum and cells. Thus, available information suggests that both cellular hypersensitivity and immune complex diseases are involved in this model of HP.

Experimental studies of bagassosis have been carried out by many workers by exposing rabbits and rats to aerosols of bagasse extracts (33,38,85,90). They monitored pulmonary pathology, precipitin production, and lymphocyte transformation of lymph node cells. Prolonged exposure was found to lead to interstitial pneumonitis with lymphocytic exudates and circulating precipitin, but no significant blast transformation of lymph node cells was observed.

Animal models of farmer's lung disease using antigens of *Micropolyspora faeni* have been produced in rats, rabbits, guinea pigs, and calves (38,75,85,100). By injecting *M. faeni* intratracheally, an HP model was produced in rabbits (33,90). These workers observed chronic inflammation in the lung, circulating precipitating antibodies, and cell-mediated immunity in the bronchoalveolar cells. It was noted that the deposition of antigen directly into the lungs leads to the induction of cell-mediated hypersensitivity. The disease could also be transferred by transfer of cells alone. On the other hand, Wilkie produced HP in guinea pigs, and he could only transfer the disease by serum and cells (100). He also found that both immune complex and cell-mediated immunity are active in the development of HP in a calf model. Using aerosols of *Thermoactinomyces candidus* spores, Hollick et al. (36) produced HP in guinea pigs. Although the peripheral cell-mediated immunologic response was negative, they demonstrated macrophage MIF in the alveolar cells after the second exposure of spores (36).

No single mechanism of defined immunologic reactivity appears to explain the pathogenesis of HP totally. The combination of both types III and IV injury produce more intense inflammation than either alone. This suggests that types III and IV immunologic injuries may be complementary. Although production of "pure" pulmonary immune complex and "pure" pulmonary delayed hypersensitivity has been reported, the present consensus favors the hypothesis of a mixed immune reaction for the actual process.

CLINICAL ASPECTS

With increased awareness of the disease, more and more cases of environmental antigen-induced pneumonitides are being reported. The number of fungal and actinomycetes species associated with the disease has also increased considerably. It should be mentioned, however, that several species of fungi and actinomycetes have been isolated from the environment in which the patient developed HP; not all were directly implicated in the pathogenesis of the pneumonitides. The antigens may be simple chemical molecules or may be large particles, such as fungus spores or dust. Particles smaller than 5 μm can be easily inhaled and deposited on the alveolar walls. Soluble antigenic materials from larger spores and other particles can also effectively sensitize the lungs and cause disease. A major requirement is a very high concentration of the antigens in the inhaled air. The following discussion is of the most common types of HP caused by fungi and actinomycetes.

Farmer's Lung

Farmer's lung results from the inhalation of mouldy hay dust contaminated with fungal spores and actinomycetes. Moist hay, during stacking, promotes the growth of bacteria and fungi. This permits fermentation of the substrate and results in an elevated temperature, providing an ideal environment for luxuriant growth of thermophilic actinomycetes. When hay dust containing fungi and actinomycetes is inhaled, susceptible individuals may develop HP. A farmer working in a mouldy atmosphere can inhale and retain in his lungs as many as 750,000 spores per minute. Approximately 98% of all spores are actinomycetes spores. Cattle also develop a similar condition during the winter months when they are confined to the stable area and are exposed to mouldy hay dust (100).

Although farmer's lung was described by Campbell (12) in 1932, the exact cause of this disease was not clearly established until recently. It has been suggested that farmer's lung may result from a hypersensitivity to moulds or to the products of moulds that occur in a wide variety of organic material (17,30). The symptoms have been described as shortness of breath, cough, fever, chills, weight loss, and hemoptysis associated with an acute granulomatous interstitial pneumonitis (17). The organisms implicated are *Micropolyspora faeni, Thermoactinomyces vulgaris, T. candidus, Saccharomonospora viridis, Aspergillus fumigatus,* and *A. flavus* (52,71,72,101). Precipitins to the sensitizing antigens usually will be present in the sera of patients. Approximately 20% of normal individuals exposed to these antigens, however, may also show antibodies.

Bagassosis

Bagasse, the dried sugar cane fiber, promotes the growth of *T. sacchari, T. vulgaris* and *T. candidus* (60,69). Bagassosis is an acute respiratory disease similar to farmer's lung that primarily affects workers after the inhalation of actinomycetes spores. This disease has been reported in most countries in which bagasse is processed.

Mushroom Worker's Lung

This hypersensitivity lung disease was first reported by Bringhurst and associates (8). Since then many more cases have been reported (14,39,83,91). Mushroom compost, a pasteurized mixture of horse excreta and vegetable matter, harbors a large number of thermophilic actinomycetes during spawning of the compost. Upon inhalation of these organisms susceptible individuals may develop characteristic HP. Reports indicate that patients develop antibodies against *Micropolyspora faeni, Thermoactinomyces vulgaris,* and *Aspergillus fumigatus,* and occasionally against compost extract and mushroom spores.

Ventilation Systems Pneumonitis

Thermophilic actinomycetes belonging to the species *Micropolyspora faeni, Thermoactinomyces vulgaris, T. candidus,* and *Saccharomonospora viridis* have been found in home and office environments (2,6,24,25,45,58). The inhalation of thermophilic actinomycetes from contaminated forced-air heating and humidification systems and air conditioners may result in the development of HP (24,25,45,58). Occasionally patients do not show antibody to any of these organisms, but may develop precipitins against humidifier water. The disease can be reproduced by challenging the patients with water from the humidifier, indicating that as yet unidentified organic antigens may be present in this material.

Maple Bark Disease

Towey et al. (94) reported 35 patients with diffuse pulmonary disease that they attributed to peeling bark from maple logs. The illness was characterized by cough, shortness of breath, fever, night sweats, and substernal pain. Emanuel et al. (20,21) reported additional cases and attributed the disease to the inhalation of spores of *Cryptostroma corticale* growing on the bark of maple logs. They also demonstrated the spores in lung biopsies and found specific antibody in the sera. The skin test with a pyridine extract of the organism produced both immediate- and delayed-type reactions. The disease was reproduced in guinea pigs, and the skin reactivity was found to be transferrable by cells but not by serum (93). The patients improved dramatically when removed from the contaminated environment.

Woodworker's Lung

A delayed hypersensitivity lung reaction characteristic of allergic alveolitis has been described among woodworkers. This may be due either to the inhalation of wood dust or to fungi contaminating the dust. Precipitating antibodies against extracts of wood dust and fungi were detected in these patients (65,89). In HP due to the inhalation of cork dust (suberosis), the serum showed precipitins against *Penicillium frequentans* (3,4). In sequoiosis (HP due to inhalation of mouldy dust from redwood trees), the commonly implicated organisms are *Graphium* spp. and *Aureobasidium pullulans;* occasionally, however, patients may show antibody to *Trichoderma* spp. (16).

Malt Worker's Lung

Malt worker's lung is an occupational lung disease seen in the malt industry. Germinating barley is heavily contaminated with *Aspergillus clavatus,* and this fungus has been recovered from the sputum of patients with this disease. Antibodies against *A. clavatus* and occasionally against *A. fumigatus* and *Rhizopus* have been detected in the sera of patients (13).

Other Occupational Diseases

Cheese washer's disease *(Penicillium casei)* and sauna taker's disease *(Aureobasidium pullulans)* are also induced by the inhalation of fungi (63,87). The inhalation of spores from puffballs *(Lycoperdon pyriforme* and *L. gemmatum)* are responsible for lycoperdonosis, a form of HP (92). Recently, *Streptomyces albus, S. olivaceous, Penicillium* spp., and *Scopulariopsis* spp. have been implicated in the etiology of HP (7,31,37,88).

Other well-documented forms of allergic alveolitis associated with occupational involvement include bird fancier's lung (23,26,77), furrier's lung, coffee worker's lung, vineyard sprayer's lung, and putuitary snuffer's lung (32,74,96,97). Pigeon breeder's disease is another well-studied form of HP resulting from the inhalation of pigeon excreta and other materials by individuals breeding or raising pigeons (23,26,27).

CLINICAL CHARACTERISTICS OF HP

The clinical features of HP depend on the immunologic response of the patient, antigenicity of organic dust, and intensity and frequency of the patient's exposure. The characteristics of the illness are generally similar, regardless of the offending antigen. Clinical manifestations, however, differ in the acute and chronic stages of the disease.

Acute Form

This common and most easily distinguishable form of HP is the result of intermittent exposure to an antigenic source. The patient becomes sensitized following exposure and subsequently develops symptoms of dyspnea, cough, fever, chills, malaise, and myalgia, which occur 4 to 8 hours after inhalation of the antigen (72). In acute HP the symptoms may persist for up to 12 hours, and the patient may recover spontaneously. Symptoms usually reappear each time the patient is exposed to the antigen(s). The degree and extent of the attack appear to depend on the amount of exposure and the sensitivity of the patient. Severe attacks may be associated with anorexia and weight loss; symptoms suggest an acute but limited viral infection, such as influenza.

Physical examination of the patient in an acute episode reveals an acutely ill, sometimes toxic-appearing and dyspneic patient. Symptoms disappear within 12 to 18 hours after a single exposure to antigens, and the patient remains asymptomatic between attacks. During the acute episode a leukocytosis as high as 25,000 cells with a marked shift to the left may be seen. Eosinophilia is variable and may be as high as 10%. These abnormalities also return to normal shortly after the acute attack subsides.

Levels of IgG, IgA, and IgM are frequently elevated in the serum of sensitive patients, but IgE levels are usually within normal limits unless the patients have, in addition to HP, an atopic condition.

During an acute attack pulmonary function studies reveal a decrease in vital and diffusing capacity. Some patients demonstrate diminished expiratory flow rates and decreased 1-second forced expiratory volumes. Chest x rays may be within normal limits or may demonstrate peripheral bibasilar infiltrations, coarsening of bronchovascular markings, fine, sharp modulations, and reticulation or honeycombing throughout the lung parenchyma, consistent with an interstitial process. There are, however, no specific features distinguishing HP from the numerous pulmonary diseases with interstitial involvement.

Both pulmonary function values and x-ray abnormalities return to normal with avoidance of exposure, provided irreversible damage to the lung parenchyma does not occur from repeated inhalation of antigens. The time involved in reversing these abnormalities usually depends on the extent and severity of the episodes. Avoidance for several weeks to a few months is generally sufficient. Corticosteroids have sometimes been used as additional therapy to reverse the detected abnormality.

Chronic Form

In some cases of HP irreversible damage to the lung occurs. This may be due to the severe immunologic reactions within the lung caused by intense and intermittent exposure to the antigens, or to mild continuous and prolonged exposure to small amounts of antigen. Fibrosis with irreversible pulmonary insufficiency can be seen in patients inhaling antigenic dusts for long periods of time. These patients may develop ventilatory impairment, diffusion defects, and "stiff" lungs that do not respond to prolonged avoidance or corticosteroid therapy. Lung biopsies from these patients show interstitial fibrosis with granuloma formation and alveolar walls thickened by infiltration with lymphocytes, plasma cells, and some eosinophils.

In a few patients with chronic HP, pulmonary function tests show elevation of their residual volume, diminished flow rates, and loss of pulmonary elasticity; these features are suggestive of emphysema. Examination of lung biopsies from such patients has revealed obstructive bronchiolitis with destruction of alveolar air sacs. These pulmonary changes are probably the result of recurrent inflammation occurring when inhaled antigen(s) interact with the precipitating antibody and/or with sensitized lymphocytes present in the lung. The inflammatory process leads to chronic bronchitis and subsequent obstructive airway disease. The chronically ill patients usually do not respond to therapy with bronchodilators, corticosteroids, or prolonged avoidance of exposure.

DIAGNOSIS

Clinical Findings

A diagnosis of HP can usually be made by appropriate environmental history and laboratory and serologic studies, and by a trial of avoidance and re-exposure to the suspected environmental antigen. The more insidious form of the disease may be difficult to diagnose, and a lung biopsy may be necessary to establish a diagnosis (Fig. 13–1). Radiologic and pulmonary physiologic examinations may provide additional information.

Confirmation of the diagnosis may be made by cautiously exposing patients to suspected antigens and carefully observing clinical reaction. The patient is first made asymptomatic by complete avoidance of the offending antigen for a period of several months, after which the patient is exposed to the environment and then observed for the next 8 hours for characteristic signs and symptoms. It may also be necessary to ascertain the specific etiologic agents by having the patient inhale the specific purified antigens directly.

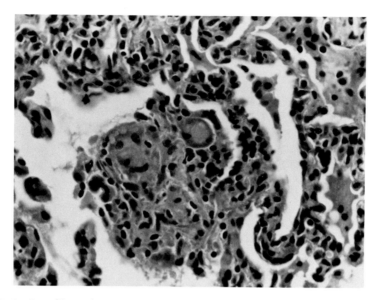

Fig. 13–1. Lung biopsy from patient with hypersensitivity pneumonitis showing pulmonary infiltrates and granuloma. (H & E × 450)

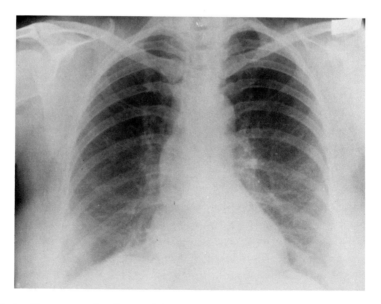

Fig. 13–2. Chest x-ray of patient with hypersensitivity pneumonitis showing diffuse interstitial infiltrates.

Radiologic Findings

X-ray findings vary from a mild interstitial infiltrate to abnormalities similar to those seen in acute pulmonary edema and in interstitial pneumonitis (Fig. 13–2). During acute attacks soft, patchy, ill-defined parenchymal densities that tend to coalesce may be seen in all lung fields. In chronic phases of the disease the lung may present the appearance of a fibrotic process with diffuse interstitial disease. The x-ray picture is usually nonspecific, and diagnosis thus depends on other features.

Physiologic Findings

Pulmonary function abnormalities vary in HP. In the acute form a primary restrictive abnormality is characteristic, although some degree of small airway disease and airway obstruction may be present. An increase in the residual volume and in the ratio of the residual volume to total lung capacity suggest the presence of small airway disease and regional air trapping. The diffusion capacity is frequently abnormal in the early part of the disease, and diminishes further with advancing severity. In the chronic form of the disease, with severe fibrosis and destruction of lung tissue, findings typical of pulmonary fibrosis, severe pulmonary emphysema with respiratory insufficiency, or both are seen.

TREATMENT

Treatment of HP depends on identification of the antigen involved. The most important therapeutic measure in HP should be avoidance of the offending antigen. Patients should use masks with filters capable of removing the dust; also, appropriate ventilation of the working area and air exchangers should be provided. Occasionally it may be essential to change the occupation. Drug therapy may be indicated in acute cases in which patients cannot avoid exposure. Corticosteroids are frequently useful in severe, acute cases; these drugs rapidly reverse the symptoms and x-ray findings. If lung damage has not progressed to the fibrotic stage, avoidance or drug therapy may result in reversal of clinical abnormalities. With irreversible lung damage, however, appropriate therapy for progressive pulmonary insufficiency may be necessary. Thus, the early detection of clinical illness, identification of the offending antigen in the environment, and institution of appropriate therapy are most important in the treatment of HP.

MICROBIOLOGIC ASPECTS

The patient's environment, suspected of containing the offending antigen, should be studied for the presence of microorganisms. Various dusts, contaminated water, and other materials from the home or working area of the patient should be cultured in appropriate media for the isolation and identification of fungi and Actinomycetes.

Isolation and Identification of Fungi

Samples are cultured in appropriate isolation media and incubated at 27° C (room temperature) and at 37° C. Liquid samples should be centrifuged at 3000 rpm for 10 minutes and the sediment suspended in sterile saline solution; solid samples are suspended in sterile saline solution (100 mg/ml) and then serial dilutions are made. One half ml of the diluted suspension is spread onto Sabouraud's dextrose agar plates (in duplicate) containing 40 units of penicillin and 20 µg of streptomycin per ml of medium. One set of the inoculated plates is then incubated at 27° C (room temperature) and the second is incubated at 37° C. Cultures are examined every day for 2 weeks. Colonies appearing on the plates can be subcultured on a fresh medium composed of Sabouraud's dextrose agar without antibiotics. The isolated fungi are identified by colonial and by microscopic morphology. When necessary, antigens are extracted from these fungi for serologic studies and for testing inhalation responses in patients.

Isolation and Identification of Thermophilic Actinomycetes

Actinomycetes are higher bacteria that resemble fungi in their colonial morphology. Many actinomycetes also produce spores resembling those of fungi. The most important of several actinomycetes species implicated in HP are those belonging to the three thermophilic genera: *Micropolyspora*, *Thermoactinomyces* and *Saccharomonospora* (45,52,58,60). Occasional reports have implicated *Streptomyces albus* and *S. olivaceous* (7,37,88). These organisms are capable of growing at elevated temperatures such as found in mouldy hay, mushroom compost, bagasse, and other vegetable matter, or in the heating and humidification systems of buildings (2,6,10,17,24,25,72,83).

When possible, specimens should be collected in sterile containers. Dust samples may be obtained by use of a vacuum cleaner and then transferred to sterile containers. Scrapings, swabs, and liquid samples are collected aseptically. Specimens should be processed as soon as possible or should be refrigerated. As in fungal isolation, samples should be suspended in sterile saline solution and appropriate dilutions should be plated onto trypticase soy agar (TSA). Novobiocin is added to the medium (30 µg/ml of medium) for controlling bacterial growth, and is supplemented with casein hydrolysate. Liquid samples should be centrifuged at 3000 rpm for 10 minutes and the deposit suspended in sterile saline solution and cultured as above. All inoculated plates should be sealed or stored in plastic bags to prevent evaporation, and incubated at 55° C for up to 1 week. Plates should be examined every day for appearance of colonies.

TABLE 13–3. Characteristics used in the Identification of Thermophilic Actinomycetes

Characteristic	Micropolyspora faeni	Saccharomonospora viridis	Thermoactinomyces candidus	T. sacchari	T. vulgaris	T. intermedius	T. dichotomica
Growth	Slow	Slow	Fast	Fast	Fast	Fast	Slow
Spores	Chain	Single (?)	Single	Single	Single	Single	Single
Aerial mycelium	+	+	+	–	+	+	+
Decomposition of:							
Casein	–	+	+	+	+	+	+
Tyrosine	+	–	–	–	+	+	–
Xanthine	+	–	–	–	–	–	–
Starch	–	+	+	+	+	–	+
Esculin	–	–	+	+	–	+	–
Arbutin	–	–	+	–	–	+	–

(Adapted from Kurup [44].)

Isolated organisms may be identified by their colonial morphology, microscopic features, physiologic characteristics, chemical composition, and immunologic characteristics (Table 13–3) (41,45,52,57,58).

Micropolyspora faeni

This organism grows slowly in most isolation media except in media containing novobiocin (medium without novobiocin should be used for isolating this organism). In the primary isolates, colonies are light yellow in color and attain a size of 2 to 4 mm in diameter (Fig. 13–3); they frequently appear similar to *Nocardia* colonies. Although slow growing, *M. faeni* can grow at 37° C, at which temperature the colony is morphologically indistinguishable from *Nocardia* spp. With prolonged incubation, a white tuft-like growth appears on the surface of the colonies. Microscopic examination of the culture shows aerial and substrate thin, branching hyphae, with chains of smooth round spores on both types. Spores are 0.8 to 1.2 μm in diameter, produced in chains of 5 to 15 units (Fig. 13–4).

The cell wall is composed of arabinose, galactose, and mesodiaminopimelic acid, indicating that the cell wall is of type IV, similar to that of *Nocardia* spp. *Micropolyspora faeni* phage specifically lyses this organism (52,57). Biochemical characteristics include hydrolysis of tyrosine and xanthine and failure to hydrolyze casein (Table 13–3). All other species

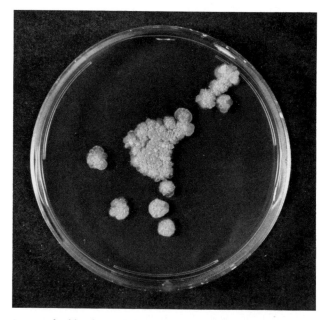

Fig. 13–3. One-week-old colony of *Micropolyspora faeni* grown on trypticase soy agar at 55° C.

Fig. 13–4. Gram-stained smear of *Micropolyspora faeni* showing chains of spores and thin hyphae. (× 1000)

included in this genus do not conform to the description of the genus, except *M. rectivirgula*. It is possible that *M. rectivirgula* and *M. faeni* are synonyms (43).

Saccharomonospora viridis

This organism was previously classified under *Thermomonospora* due to the production of spores only on aerial hyphae. Results of recent studies, however, indicate that the cell wall is of type IV, while *Thermomonospora* possess a type III cell wall (42). The colonies grow slowly in most media, initially appearing white but turning to bluish-green when spores are produced (Fig. 13–5). Casein hydrolysate medium supports the production of spores. The spores are oval, 1.5 × 1.0 μm in size, and produced on dichotomously branched sporophores on aerial hyphae only (Fig. 13–6). The spores do not retain malachite green when spore stain is used (43).

The ideal temperature for growth is between 45 and 50° C. The organism hydrolyzes casein and starch but is unable to decompose tyrosine, xanthine, or hypoxanthine. Like *Micropolyspora faeni*, *S. viridis* is also highly sensitive to novobiocin. This organism is frequently isolated from mouldy hay, corn, and other vegetable matter, and from indoor heating systems (43,57).

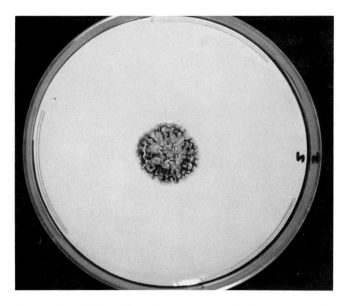

Fig. 13–5. One-week-old colony of *Saccharomonospora viridis* grown on casein hydrolysate agar at 45° C.

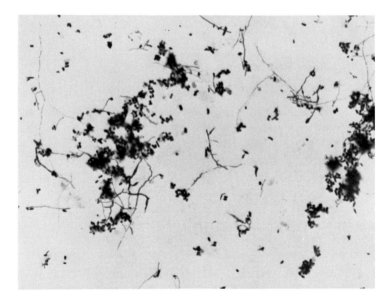

Fig. 13–6. Gram-stained smear showing oval spores of *Saccharomonospora viridis* produced only on aerial hyphae (× 450)

Thermoactinomyces spp.

Five of the seven species included in this genus, *T. candidus, T. sacchari, T. vulgaris, T. intermedius,* and *T. dichotomica,* are frequently isolated from the environment (44,50,52,68). The first three species are associated with HP. The role of the latter two in HP is unknown, although they share several antigens with other species. All these species are resistant to novobiocin.

Thermoactinomyces candidus and *T. vulgaris* grow well in all common laboratory media. These species are fast-growing and produce fluffy or granular white colonies (Fig. 13–7). Both species grow at 55 to 60° C and usually fail to grow at 37° C or less. Spores are produced on both aerial and substrate hyphae and are either produced on small sporophores ranging from 1 to 3 μm in size or directly on the hyphae (Fig. 13–8). The spores are typical endospores, contain dipicolinic acid, and are thermoresistant. Casein is hydrolyzed by all species of *Thermoactinomyces. T. vulgaris* hydrolyzes tyrosine and starch but not esculin or arbutin, while *T. candidus* hydrolyzes arbutin and esculin and is lysed by specific phages that do not lyse any of the other species. Antigenic differences are also noted between the two species.

Thermoactinomyces sacchari grows very slowly on half-strength nutrient agar and on casein hydrolysate agar. The colonies are usually transparent and grow well at 50° C. The aerial hyphae are lysed in 3 to 4 days. Other characteristics are the same as those of *T. vulgaris* and *T. candidus,* except that *T. sacchari* hydrolyzes casein starch and esculin but not tyrosine or arbutin. Spores (endospores) are always produced on slender and long sporophores measuring 2 to 3 μm.

Thermoactinomyces dichotomica produces slow-growing yellowish colonies. This species demonstrates all the characteristics of the genus: hydrolyzing casein, starch, and gelatin, producing endospores, and growing at 55 to 60° C. *T. intermedius* morphologically resembles *T. candidus* and *T. vulgaris,* but fails to hydrolyze starch. It can decompose casein and tyrosine. Immunologically this species shares antigens with *T. vulgaris* and *T. candidus.*

Thermoactinomyces candidus and *T. vulgaris* have been isolated from a wide range of substrates, including mushroom compost, mouldy hay and corn, vegetable matter, and heating and humidification systems in residential and commercial buildings (56). *T. sacchari* has been isolated from mouldy bagasse. All three species are implicated in HP; the remaining species are less common and their roles in HP have not been documented.

SEROLOGIC TESTS

All forms of HP are characterized by the presence of circulating antibodies against the respective antigen. Dust extract, water samples, or

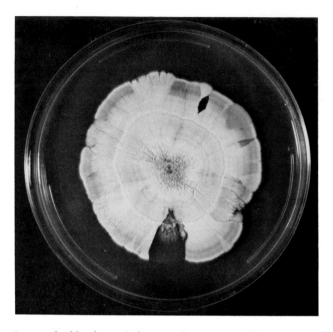

Fig. 13–7.   One-week-old colony of *Thermoactinomyces candidus* grown on trypticase soy agar at 55° C.

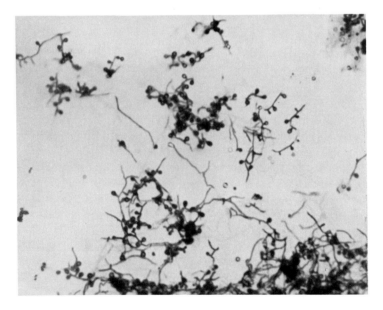

Fig. 13–8.   Gram-stained smear showing single spores produced by *Thermoactinomyces candidus* on short sporophores or directly on the hyphae. (× 1000)

antigens from cultured organisms may be used to detect precipitating antibodies in the patient's serum. Several methods such as complement fixation, hemagglutination, indirect immunofluorescence, immunoelectrophoresis, and crossed immunoelectrophoresis have been used for the detection of antibodies (19,29,51,67,72,73,95,99). The agar gel double diffusion method is the most widely used because of its simplicity.

Antigen Preparations

Crude Extracts from the Sample

Dust and other solid samples (500 mg) are extracted overnight in 0.5% phenol-saline solution (0.85% NaCl). Following centrifugation, the supernatant may be used either directly or may be concentrated. Liquid samples may also be used directly or after being concentrated.

Fungal Antigens

Fungi are grown in a synthetic medium, and the culture filtrate antigens are extracted (47). AOAC broth (Difco) supplemented with 1% glucose is recommended for growing the organism. The broth (500 ml), in 2-liter flasks, should be inoculated with a suspension of spores (10^8–10^9 spores per 500 ml) and incubated at either 27 or 37° C, depending on the requirements of the organism. Cultures should be incubated for 3 weeks and then treated with 0.5% formalin and allowed to stand at 4° C for 2 days, at which time their viability can be ascertained. If necessary, additional formalin can be added and tested as before. The broth, separated by filtration, can then be dialyzed against deionized water at 4° C for 4 to 5 days, filtered through a membrane filter (0.45 μm), and freeze-dried. The antigens can be diluted to 10 mg/ml and used for the agar gel diffusion method (48).

Thermophilic Actinomycetes Antigens

Most thermophilic actinomycetes grow well in complex media but fail to grow in any of the simple media. The removal of macromolecular components of the media from antigen is extremely difficult. Occasionally these components may react with the sera of patients and give a false positive reaction. These antigens are not suitable for skin testing due to the presence of impurities.

A double dialysis method has been developed to exclude media components (18,46). With this method the complex medium, trypticase soy broth (TSB), is placed inside a dialysis bag that has a molecular weight retention of 8000 daltons, and the whole bag is placed into a glycine and sodium chloride solution (glycine, 0.1 M; NaCl, 0.075 M). Following autoclaving, the system is allowed to equilibrate for 48 hours and the outer fluid is collected aseptically and inoculated with the organism. The inoculated fluid is incubated at 55° C for 2 weeks, 0.5% formalin is then added, and the system is processed as described for fungal antigens. *Thermoactinomyces vulgaris, T. candidus,* and *S. viridis* may be grown in this manner. *T. sacchari* may be cultured in TSB and starch dialysate and incubated at 50° C for two weeks. Antigens should be reconstituted at 10 to 30 mg/ml for the immunodiffusion test.

A synthetic medium has been devised to grow *Micropolyspora faeni* (49,53,55). AOAC broth (Difco), supplementedwith 1% lactose and 0.01% spermidine, is inoculated with M. *faeni,* and the inoculated broth is incubated on a shaker at 55° C for 5 to 7 days. After adding 0.5% formalin, the culture filtrates are separated as for other antigens. A concentration of 5 mg/ml is usually used by agar gel diffusion to detect precipitins in patients' sera.

Agar Gel Double Diffusion

A modification of the microimmunodiffusion technique of Wadsworth can be used (54,98). Plastic templates are placed on microscopic slides and layered with 0.8% agar in 0.1 M barbital buffer (pH 8.6). The center wells are filled with antigens and the peripheral wells are filled with sera from patients. These are allowed to diffuse in a moist chamber for 48 hours at 27° C. Controls of known positive and negative sera are included with each set. The templates are then removed and the slides are washed in physiologic saline solution (0.15 M NaCl) for 24 hours to remove the unreacted antigen and sera. After final washing in distilled

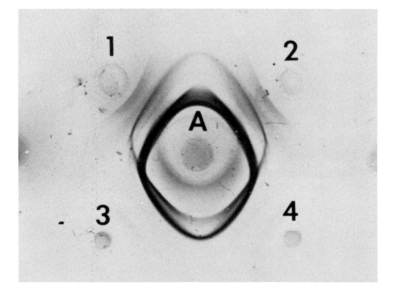

Fig. 13–9. Agar gel diffusion plate stained by Coomassie blue stain. Serum of patient with hypersensitivity pneumonitis against antigens extracted from organism isolated from patient's own environment (1, 2) and against antigens from a standard *Thermoactinomyces candidus* strain (T-106) (3, 4). A, antigens.

Fig. 13–10. Antigen-antibody crossed immunoelectrophoresis of *Micropolyspora faeni* antigens against a farmer's lung patient serum.

water the slides are stained by Coomassie blue R-250 (0.2% in metha-nol:acetic acid:water, 4.5:1:4.5). The precipitin bands are observed and evaluated in comparison with the controls (Fig. 13–9). Farmer's lung sera frequently show several precipitin bands when studied by antigen-antibody crossed immunoelectrophoresis (Fig. 13–10).

COMMENTS

The immunopathogenesis of HP, including the role of fungal antigens, is quite complex; the presence of several factors is probably required to produce symptomatic disease. Exposure must occur by inhalation of antigens that most likely possess unique characteristics, such as non-digestibility and/or adjuvant or complement-activating properties. The qualitative and quantitative requirement of exposure may be important but are not defined at present. The exposed individual must also be susceptible to development of the disease. This may involve a genetic predisposition or an acquired susceptibility in terms of immunologic or target organ sensitivity to inhaled antigens. Further studies of the unique characteristics of antigens associated with HP and necessary exposure circumstances are needed. Studies of predisposing factors and immunologic response, with emphasis on quantitative parameters, as well as studies of the combination of reactivity and local pulmonary immunologic factors are also required for better understanding of the disease.

More information on the spectrum of pulmonary reaction to inhaled antigen is needed to define the mechanisms of HP in patients. Increased knowledge of these mechanisms will also increase our understanding of the basic immune mechanism in the lung. The role of antigens and the sequential events in the genesis of the disease need to be elucidated for a fuller understanding of the disease.

It is also essential to determine how the host responds to the inhaled antigen, and how the antigen is broken down and removed from the patient's system. Knowledge of the role of immune complexes and cell-mediated immune reponses, and the mediators responsible for tissue damage, would be of considerable value in designing proper preventive measures and more efficient treatment.

REFERENCES

1. Allen, D.H., et al.: Familial hypersensitivity pneumonitis. Am. J. Med., **59**:505, 1975.
2. Arnow, P.M., et al.: Early detection of hypersensitivity pneumonitis in office workers. Am. J. Med., **64**:236, 1978.
3. Avila, R., and Lacey, J.: The role of *Penicillium frequentans* in suberosis. (Respiratory disease in workers in the cork industry.) Clin. Allergy, **4**:109, 1974.
4. Avila, R., and Villar, T.G.: Suberosis, respiratory disease in cork workers. Lancet, **1**:620, 1968.
5. Bach, F.H., and Van Rood, J.J.: The major histocompatibility complex—genetics and biology. N. Engl. J. Med., **295**:927, 1976.
6. Banaszak, E.F., Thiede, W.H., and Fink, J.N.: Hypersensitivity pneumonitis due to contamination of an air conditioner. N. Engl. J. Med., **283**:271, 1970.
7. Blackburn, C.R.B., and Green, W.: Precipitins against extracts of thatched roofs in the sera of New Guinea natives with chronic lung disease. Lancet, **2**:1396, 1966.
8. Bringhurst, L.S., Byrne, R.N., and Gershon-Cohen, J.: Respiratory disease of mushroom workers: Farmer's lung. JAMA., **171**:15, 1959.
9. Buechner, H.A.: Bagassosis: Peculiarities of its geographical pattern and report of the first case from Peru and Puerto Rico. JAMA, **174**:1237, 1960.
10. Buechner, H.A., et al.: Bagassosis—a review with further historical data, studies of pulmonary function and results of adrenal steroid therapy. Am. J. Med., **25**:234, 1958.
11. Caldwell, J.R., et al.: Immunologic mechanism in hypersensitivity pneumonitis. J. Allergy Clin. Immunol., **52**:225, 1973.
12. Campbell, J.M.: Acute symptoms following work with hay. Br. Med. J., **11**:1143, 1932.
13. Channell, S., et al.: Allergic alveolitis in malt workers. Q. J. Med., **38**:351, 1969.
14. Chan-Yeung, M., Grzybowski, S., and Schonell, M.E.: Mushroom worker's lung. Am. Rev. Resp. Dis., **105**:819, 1972.
15. Christensen, L.T., Schmidt, C.D., and Robins, L.: Pigeon breeder's disease—a prevalence study and review. Clin. Allergy, **5**:417, 1975.
16. Cohen, H.I., et al.: Sequoiosis, a granulomatous pneumonitis associated with redwood sawdust inhalation. Am. J. Med., **43**:785, 1967.
17. Dickie, H.A., and Rankin, J.: Farmer's lung. An acute granulomatous interstitial pneumonitis occurring in agricultural workers. JAMA, **167**:1069, 1958.
18. Edwards, J.H.: The double dialysis method of producing farmer's lung antigens. J. Lab. Clin. Med., **79**:683, 1972.
19. Edwards, J.H., Baker, J.T., and Davies, D.H.: Precipitin test negative in farmer's lung—activation of the alternative pathway of complement by mouldy hay dusts. Clin. Allergy, **4**:379, 1974.
20. Emanuel, D.A., Lawton, B.R., and Wenzel, F.J.: Maple bark disease. Pneumonitis due to *Coniosporium corticale*. N. Engl. J. Med., **266**:333, 1962.
21. Emanuel, D.A., Wenzel, F.J., and Lawton, B.R.: Pneumonitis due to *Cryptostroma corticale* (maple bark disease). N. Engl. J. Med., **274**:1413, 1966.

22. Fink, J.N.: Hypersensitivity pneumonitis due to organic dusts. Clin. Notes Resp. Dis., **13**:3, 1974.
23. Fink, J.N., et al.: Pigeon breeder's disease—a clinical study of hypersensitivity pneumonitis. Ann. Intern. Med., **68**:1205, 1968.
24. Fink, J.N., et al.: Interstitial pneumonitis due to hypersensitivity to an organism contaminating a heating system. Ann. Intern. Med., **74**:80, 1971.
25. Fink, J.N., et al.: Interstitial lung disease due to contamination of forced air systems. Ann. Intern. Med., **84**:406, 1976.
26. Fink, J.N., Barboriak, J.J., and Sosman, A.J.: Immunological study of pigeon breeder's disease. J. Allergy, **39**:214, 1967.
27. Fink, J.N., Moore, V.L., and Barboriak, J.J.: Cell-mediated hypersensitivity in pigeon breeders. Int. Arch. Allergy Appl. Immunol., **49**:831, 1975.
28. Flaherty, D.K., et al.: HL-A-8 in farmer's lung. Lancet, **2**:507, 1975.
29. Fletcher, S.M., Rondle, C.J.M., and Murray, I.G.: The extracellular antigens of *Micropolyspora faeni:* Their significance in farmer's lung disease. J. Hyg. (Camb.), **68**:401, 1970.
30. Fuller, C.J.: Farmer's lung: A review of present knowledge. Thorax, **8**:59, 1953.
31. Grieble, H.G., et al.: Scopulariopsis and hypersensitivity pneumonitis in an addict. Ann. Intern. Med., **83**:326, 1975.
32. Harper, L.O., et al.: Allergic alveolitis due to pituitary snuff. Ann. Intern. Med., **73**:581, 1970.
33. Harris,J.O., Bice, D., and Salvaggio, J.E.: Cellular and humoral bronchopulmonary immune response of rabbits immunized with thermophilic actinomycetes antigen. Am. Rev. Resp. Dis., **114**:29, 1976.
34. Hensley, G.T., et al.: Lung biopsies of pigeon breeder's disease. Arch. Pathol., **87**:572, 1969.
35. Hensley, G.T., Fink, J.N., and Barboriak, J.J.: Hypersensitivity pneumonitis in the monkey. Arch. Pathol., **97**:33, 1974.
36. Hollick, G.E., Hall, N.K., and Larsh, H.W.: Peripheral and alveolar reponse in guinea pigs to an aerosol exposure of *Thermoactinomyces candidus* spores. Mykosen, **23**:120, 1979.
37. Kagan, S.L., et al.: *Streptomyces albus:* A new cause of hypersensitivity pneumonitis. J. Allergy Clin. Immunol., **68**:295, 1981.
38. Kawai, T., et al.: Alveolar macrophage migration inhibition in animals immunized with thermophilic actinomycetes antigen. Clin. Exp. Immunol., **15**:123, 1973.
39. Kleyn, J.G., Johnson, W.M., and Wetzler, T.F.: Microbial aerosols and actinomycetes in etiological considerations of mushroom worker's lung. Appl. Environ. Microbiol., **41**:1454, 1981.
40. Kobayashi, M.: Antigens in moldy hay as the cause of farmer's lung. Proc. Soc. Exp. Bio. Med., **113**:472, 1963.
41. Kurup, V.P.: Identification of thermophilic actinomycetes. J. Allergy Clin. Immunol., **61**:232, 1978.
42. Kurup, V.P.: Characterization of some members of the genus *Thermomonospora.* Curr. Microbiol., **2**:267, 1979.
43. Kurup, V.P.: Taxonomic study of some members of *Micropolyspora* and *Saccharomonospora.* Microbiologica, **4**:249, 1981.
44. Kurup, V.P.: Thermophilic actinomycetes associated with hypersensitivity pneumonitis. Science-Ciencia, **8**:5, 1981.
45. Kurup, V.P., et al.: *Thermoactinomyces candidus,* a new species of thermophilic actinomycetes. Int. J. Syst. Bacteriol., **25**:150, 1975.
46. Kurup, V.P., et al.: Immunologic cross-reaction among thermophilic actinomycetes associated with hypersensitivity pneumonitis. J. Allergy Clin. Immunol., **57**:417, 1976.
47. Kurup, V.P., et al.: Antigenic variability of *Aspergillus fumigatus* strains. Microbios, **19**:191, 1978.
48. Kurup, V.P., et al.: The detection of circulating antibodies against antigens from three strains of *Aspergillus fumigatus.* Mykosen, **23**:368, 1980.
49. Kurup, V.P., et al.: Characterization of *Micropolyspora faeni* antigens. Infect. Immunol., **34**:508,1981.
50. Kurup, V.P., Barboriak, J.J., and Fink, J.N.: Comments on taxonomy of some *Thermoactinomyces species.* Biol. Actinomycetes Related Organisms, **12**:53, 1977.

51. Kurup, V.P., Barboriak, J.J., and Fink, J.N.: Indirect immunofluorescent detection of antibodies against thermophilic actinomycetes in patients with hypersensitivity pneumonitis. J. Lab. Clin. Med., **89**:533, 1977.

52. Kurup, V.P., and Fink, J.N.: A scheme for the identification of thermophilic actinomycetes associated with hypersensitivity pneumonitis. J. Clin. Microbiol., **2**:55, 1975.

53. Kurup, V.P., and Fink, J.N.: Extracellular antigens of *Micropolyspora faeni* grown in synthetic medium. Infect. Immunol., **15**:608, 1977.

54. Kurup, V.P., and Fink, J.N.: Antigenic relationships among thermophilic actinomycetes. Sabouraudia, **17**:163, 1979.

55. Kurup, V.P., and Fink, J.N.: Antigens of *Micropolyspora faeni* strains. Int. Arch. Allergy Appl. Immunol., **60**:140, 1979.

56. Kurup, V.P., Fink, J.N., and Bauman, D.M.: Thermophilic actinomycetes from the environment. Mycologia, **68**:662, 1976.

57. Kurup, V.P., and Heinzen, R.J.: Isolation and characterization of actinophages of *Thermoactinomyces* and *Micropolyspora*. Can. J. Microbiol., **24**:794, 1978.

58. Kurup, V.P., Hollick, G.E., and Pagan, E.F.: *Thermoactinomyces intermedius,* a new amylase-negative thermophilic actinomycetes. Science-Ciencia, **7**:104, 1980.

59. LaBerge, D.E., and Stahmann, M.A.: Antigens from *Thermopolyspora polyspora* involved in farmer's lung. Proc. Soc. Exp. Biol. Med., **121**:463, 1966.

60. Lacy, J.: *Thermoactinomyces saccharii* sp. nov. A thermophilic actinomycete causing bagassosis. J. Gen. Microbiol., **66**:327, 1971.

61. Marx, J.J., Jr., et al.: Migration inhibition factor and farmer's lung antigens (abstract). Clin. Res., **21**:852, 1973.

62. Marx, J.J., Jr., et al.: Frequency of farmer's lung disease among farmers with antibodies to the thermophilic actinomycetes. Am. Rev. Resp. Dis., **113**(Suppl.):128, 1976.

63. Metzger, W.J., et al.: Sauna taker's disease. JAMA, **236**:2209, 1976.

64. Moore, V.L., Hensley, G.T., and Fink, J.N.: An animal model of hypersensitivity pneumonitis in rabbits. J. Clin. Invest., **56**:937, 1975.

65. Mue, S., et al.: A study of western red cedar sensitivity: Worker's allergy reaction and symptoms. Ann. Allergy, **35**:148, 1975.

66. Nicholson, D.P.: Extrinsic allergic pneumonias. Am. J. Med., **53**:131, 1972.

67. Nielsen, K.H., Parratt, D., and White, R.G.: Quantitation of antibodies to particulate antigens using a radiolabeled anti-immunoglobulin reagent. J. Immunol. Methods, **3**:301, 1973.

68. Nonomura, H., and Ohara, Y.: Distribution of actinomyces in soil. X. New genus and species of monosporic actinomycetes. J. Ferm. Technol., **49**:95, 1971.

69. Pagan, E.F., et al.: Prevalence of thermophilic Actinomycetes in sugar cane-related environments and precipitins in the population in Puerto Rico. Biol. Assoc. Med. P.R., **73**:234, 1981.

70. Patterson, R., et al.: Antibodies of different immunoglobulin classes against antigen causing farmer's lung. Am. Rev. Resp. Dis., **114**:315, 1976.

71. Patterson, R., Sommers, H., and Fink, J.N.: Farmer's lung following inhalation of *Aspergillus flavus* growing in moldy corn. Clin. Allergy, **4**:79, 1974.

72. Pepys, J.: Monographs in allergy. *In* Hypersensitivity Diseases of the Lungs due to Fungi and Organic Dusts. Vol. 4. Edited by P. Kallos, et al. Basel, S. Karger, 1969.

73. Pepys, J., Riddell, R.W., and Citron, K.M.: Precipitins against extracts of hay and fungi in the serum of patients with farmer's lung. Acta Allergy, **16**:76, 1961.

74. Pimentel, J.C.: Furrier's lung. Thorax, **25**:387, 1970.

75. Pirie, H.M., et al.: A bovine disease similar to farmer's lung: Extrinsic allergic alveolitis. Vet. Rec., **88**:346, 1971.

76. Ramazzini, B.: De Morbis Artificum Diatriba 1971. (Translated by W.C. Wright.) Chicago, University of Chicago Press, 1940.

77. Reed, C.E., Sosman, A.J., and Barbee, R.A.: Pigeon breeder's lung—a new observed interstitial pulmonary disease. JAMA, **193**:261, 1965.

78. Reyes, C.N., et al.: The histopathology of farmer's lung (60 consecutive cases). Am. J. Pathol., **66**:460, 1976.

79. Richerson, H.B.: Acute experimental hypersensitivity pneumonitis in the guinea pig. J. Lab. Clin. Med., **79**:745, 1972.

80. Richerson, H.B.: Varieties of acute immunologic damage to the rabbit lung. Ann. N.Y. Acad. Sci., **221**:340, 1974.
81. Robinsin, T.J.: Coeliac disease with farmer's lung. Br. Med. J., **1**:745, 1976.
82. Roska, A.K., et al.: Immune complex disease in guinea pigs. I. Elicitation by aerosol challenge, suppression with cobra venom factor and passive transfer with serum. Clin. Immunol. Immunopathol., **8**:213, 1977.
83. Sakula, A.: Mushroom worker's lung. Br. Med. J., **3**:708, 1967.
84. Salvaggio, J.E., et al.: Bagassosis. 1. Precipitins against extracts of crude bagasse in the serum of patients. Ann. Intern. Med., **64**:748, 1966.
85. Salvaggio, J., et al.: Experimental production of granulomatous pneumonitis. J. Allergy Clin. Immunol., **56**:364, 1975.
86. Santives, T., et al.: Immunologically induced lung disease in guinea pigs. A comparison of ovalbumin and pigeon serum as antigens. J. Allergy Clin. Immunol., **57**:582, 1976.
87. Schlueter, D.P.: "Cheese washer's disease." A new occupational hazard. Ann. Intern. Med., **78**:606, 1973.
88. Solley, G.O., and Hyatt, R.E.: Hypersensitivity pneumonitis induced by *Penicillium* species. J. Allergy Clin. Immunol., **65**:65, 1980.
89. Sosman, A.J., et al.: Hypersensitivity to wood dust. N. Engl. J. Med., **281**:977, 1969.
90. Stankus, R.P., Cashner, F.M., and Salvaggio, J.E.: Bronchopulmonary macrophage activation in the pathogenesis of hypersensitivity pneumonitis. J. Immunol., **120**:685, 1978.
91. Stewart, C.J.: Mushroom worker's lung—two outbreaks. Thorax, **29**:252, 1974.
92. Strand, R.D., Neuhauser, E.B.D., and Sornberger, C.F.: Lycoperdonosis. N. Engl. J. Med., **277**:89, 1967.
93. Tewksbury, D.A., Wenzel, F.J., and Emanuel, D.A.: An immunological study of maple bark disease. Clin. Exp. Immunol., **3**:857, 1968.
94. Towey, J.W., Sweany, H.C., and Huron, W.H.: Severe bronchial asthma due to fungus spores found in maple bark. JAMA, **99**:453, 1932.
95. Treuhaft, M.W., et al.: Characterization of precipitin response to *Micropolyspora faeni* in farmer's lung disease by quantitative immunoelectrophoresis. Am. Rev. Resp. Dis., **119**:571, 1979.
96. Villar, T.G.: Vineyard sprayer's lung. Clinical aspects. Am. Rev. Resp. Dis., **110**:545, 1974.
97. Van Toorn, D.W.: Coffee worker's lung, a new example of extrinsic allergic alveolitis. Thorax, **25**:399, 1970.
98. Wadsworth, C.: Microtemplate employing a gel chamber compared with other micro- and macrolate templates for immunodiffusion. Int. Arch. Allergy Appl. Immunol., **21**:131, 1962.
99. Wenzel, F.J., et al.: Serologic studies in farmer's lung. Precipitins to thermophilic actinomycetes. Am. Rev. Resp. Dis., **109**:464, 1974.
100. Wilkie, B.N.: Hypersensitivity pneumonitis. Experimental production in calves with antigens of *Micropolyspora faeni*. Can. J. Comp. Med., **40**:221, 1976.
101. Yocum, M.W., et al.: Extrinsic allergic alveolitis after *Aspergillus fumigatus* inhalation. Evidence of type IV immunologic pathogenesis. Am. J. Med., **61**:939, 1976.

ACKNOWLEDGMENTS

This work was supported by Asthma and Allergic Diseases Center Grant #AI 19104 from the NIH, the Veterans Administration, and the Marcus Center for Immunological and Allergy Research.

Chapter 14

MOULD ALLERGY AND CLIMATE CONDITIONS

Mark R. Sneller

Allergenic fungal species have been identified virtually everywhere that appropriate measuring devices have been established. Fungal spores have been reported to be common at 1,000 m (approximately 3,333 ft) (38,63) and have been recovered from altitudes as high as 7,000 to 10,000 ft (38). They have been found in surprisingly high concentrations in clouds (63) and in the air during almost every type of climatic condition. Some moulds, such as *Cladosporium (Hormodendrum), Alternaria, Fusarium, Helminthosporium,* and the yeasts are considered to be universal dominants (69) and, with some variation, are found in worldwide aerobiologic surveys. These fungi almost always occur seasonally. The indoor concentration of these fungi is dependent on the outdoor concentration. Other universal dominants, such as *Aspergillus* and *Penicillium,* are usually nonseasonal in occurrence and can be expected to be found on a regular basis indoors. Some moulds are more common inland rather than in coastal areas (7), and some are prevalent during foggy weather (63). Fungi have been found on rocks in the direct sunshine of the world's deserts (91) as well as in the air of considerably colder climates (25).

The relationships between climate and mould are complex, and must be examined from numerous perspectives. Early reports on mould allergy attributed the source of the allergy to mould in the patient's home or in his occupational exposure (9,15,43). In 1935 Feinberg reported that the outside air was also a significant source of mould spores (28). This gave impetus to the study of environmental conditions affecting the prevalence of moulds indoors and outdoors.

TABLE 14–1. Climatic and Climate-Related Factors Affecting Mould Allergy

NATURAL CLIMATES
  Single factors
    Barometric pressure
    Relative humidity
    Temperature
    Wind
    Positive ions
    Precipitation
    Fog
    Type of atmospheric mould
    Amount of atmospheric mould
    Local and nonlocal vegetation
  Multiple factors
    Inversion layers and stagnant conditions
    Hot wind (above body temperature)
    Cold wind (wind chill)
    Barometric pressure and relative humidity
    Temperature and relative humidity
    Rapid changes in all variables
MAN-MADE CLIMATES
  Indoor cooling devices (humidifiers, air conditioners)
  Atmospheric dust caused by automobile traffic on unpaved roadways
  Mould substrates from agriculture, imported trees and shrubs, house
      plants
  Heating ducts containing dust and spores
  Damp basements, walls, shower curtains

Table 14–1 lists the climatic factors affecting mould allergy. Natural and man-made climates each play an important role in the initiation and exacerbation of allergic symptoms due to mould. Natural climatic factors are most important in combination with one another, not only in terms of permitting or inhibiting growth of mould but in terms of distributing such mould, as well as sensitizing the human body to allergenic substances. Diverse factors such as blowing dust and spores, drying of mucous membranes, wind chill, presence of indoor cooling devices, indoor plants, and passage of storm fronts may all play a role when it comes to a particular patient exhibiting symptoms of mould allergy. Figure 14–1 illustrates the central role of climate in determining the effectiveness of fungal allergenicity. This chapter will emphasize certain points regarding climate-mould relationships in the various climatic regions of the world, as follows:

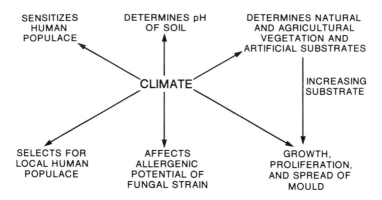

Fig. 14–1. The central role of climate in determining the effectiveness of fungal allergenicity.

1. There exists a complex series of interactions that permits some fungi to be seasonal and those same fungi to be nonseasonal.
2. Results of investigations around the world indicate that fungi are allergens, and that climate is the force that moves these allergens and sensitizes humans to their effects.
3. Outdoor climate affects indoor climate (man-made climate), and indoor climate can be a problem when it comes to the propagation of moulds.
4. Agricultural activities as well as the "natural" environment must be monitored for the presence of allergenic fungi.
5. Despite vastly different climatic regions around the world fungal occurrence can be predicted following basic laws of behavior.
6. Climate can act to precipitate diseases such as asthma, with the right weather and the right person.

OUTDOOR-INDOOR MOULD RELATIONSHIPS

Numerous climate-mould relationships emerged from a study from the agriculturally oriented Salinas Valley of Northern California (84,85). This investigation covered a 12-month period, from November 1977 through October 1978. Allergy patients (145) were given verbal and written instructions to expose a set of five gravity settling plates (Petri plates) at the same time of day for a 10-minute period. Different rooms of the patients' homes were sampled, and outdoor samples were taken by exposing similar plates near open windows.

In this investigation *Cladosporium* served as a marker organism; that is, it was a fungal model for studying factors affecting fungal prevalence in and out of doors. *Cladosporium* was chosen because it was recovered in 90 and 70% of the indoor and outdoor samples, respectively.

Figure 14–2 shows how outdoor climate affects indoor climate. An initial rise in outdoor counts occurs in the same weeks as a sharp change

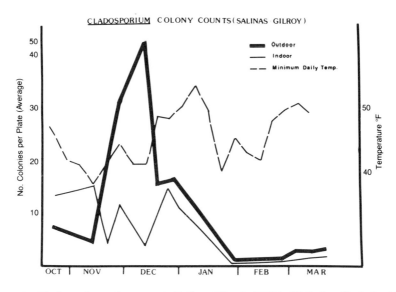

Fig. 14–2   *Cladosporium* **colony counts (Salinas-Gilroy), 1977 to 1978: the effect of outdoor climate and mould on indoor climate and mould.**

in minimum daily temperature (MDT); this is followed by a sharp rise in indoor colony counts. The number of outdoor colony-forming units (CFU) begins to fall during the first week in December, with a subsequent decrease in incidence of indoor CFU detected 1 week later. Fluctuations in frequency of indoor CFU are observed to occur during later November and December.

Increases or decreases in outdoor CFU have been shown elsewhere to cause similar changes in indoor CFU (71,89). In the study cited above, decreases in indoor CFU are more closely and consistently linked to sharp changes in MDT than to other climatologic factors. These sharp changes cause an increase in convection currents within the home and spores become airborne (35,51,77), thus yielding fewer colonies per plate. This is important to atopic individuals, because rapid changes in the aeroallergen content of the air frequently cause exacerbation of allergic symptoms (37,77).

Given the presence of initially high spore numbers, precipitation is known to cause a decrease in outdoor CFU due to washing of spores from the air (48). It is possible that this phenomenon occurs when a sharp drop in spore counts is observed in December, a period associated with heavy rainfall.

Conventionally ventilated homes are readily penetrated by naturally occurring particles, so in most cases more viable spores of a fungus have been recovered outdoors than indoors (22). Many fungi such as *Penicillium* and *Aspergillus,* however, can also be of endogenous origin, not following any recognizable seasonal or climatic pattern, and can also be of importance to the mould-sensitive patient.

EFFECT OF CLIMATE ON HOURLY SPORE FREQUENCY

In 1979 use of the Burkard sampler in Tucson enabled us to recover a number of spores and fragments that resemble those belonging to the following fungi genera: *Dematium, Bipolaris, Alternaria, Stemphylium, Seimatosporium, Leptosphaeria, Bispora, Macrosporium, Cylindrocarpon, Cercosporella, Pithomyces, Fusarium, Curvularia, Thielaviopsis, Microperra, Aspergillus, Helminthosporium, Rhizopus, Cladosporium,* smut cells, Basidiomycetes spores, and unknown species.

Although many of the above fungi are members of the Ascomycetes, tentatively identified on the basis of the appearance of their ascospores, others such as *Aspergillus* and *Rhizopus* were captured with their fruiting structures intact (Table 14–2).

Clumps of spores grow as a single colony on media, but in terms of allergenic importance the individual particles comprising the clump are significant. Over 500 *Cladosporium* conidia, most occurring in clumps or still attached to the condiophores, were observed over a 1-hour period around midnight of the first week in August 1979. Similarly, chains of three and four *Alternaria* spores were not infrequently observed during times of peak *Alternaria* incidence, and these occurred concomitant with large numbers of conidiophores identical to *Alternaria* conidiophores. These data suggest that *Alternaria* conidia, reported to be released in the early morning hours (44), begin to settle to the ground in the afternoon, and that conidiophores may follow the same pattern (44).

It is likely, however, that the *Alternaria* conidiophore is dissociated from its mycelium at a different time of the day than the conidia because they are recovered at the same time. This is due to vast differences in the physical properties of the two particle types, which affect their settling rate (37, 58). According to Stoke's law, asymmetry and roughness of the "spore" would presumably affect the rate of fall (35). The size, shape, and approximate volume of fungal spores have been described by Harsh and Allen (37).

Ingold (48) has pointed out that early morning humidity is probably important for the release of Sporobolomyces daughter cells, and that wind velocity may be important for the release of barley smut. In still other fungi, diurnal rhythms in spore liberation are associated with light, the change from darkness to light, or temperature-relative humidity relationships.

In Table 14–2 the category of "Others" primarily includes members of the Ascomycetes and some Basidiomycetes. Although in Ascomycetes a light-conditioned rhythm of spore discharge appears to be common, it is apparently rare in Basidiomycetes (48). It is therefore difficult to

TABLE 14–2. Number of Fungal Units (×10³) Counted in Tucson, Arizona, in 1979*

Fungus	Time (hours)							Percentage	
	2400	0400	0800	1200	1600	2000	Total	Overall	Range
Cladosporium	14.3	9.7	7.7	5.7	11.6	8.3	57.3	25	13–38
Alternaria	8.0	6.7	7.8	7.1	10.6	9.3	49.5	21	11–33
Smut	10.5	6.1	5.9	4.7	6.9	9.2	43.3	19	2–52
Others	8.6	8.7	8.9	8.6	7.6	9.7	52.1	22	9–44
Mycelial fragments	3.7	3.1	4.3	4.6	4.0	4.0	23.7	10	2–28
Helminthosporium	0.8	0.5	1.5	1.2	1.5	1.0	6.5	3	0–6
Hourly totals:	45.9	34.8	36.1	31.9	42.2	41.5	232.4	100	

*The total numbers of each of the fungi should be multiplied by four to determine the predicted number of fungi for the year because only 6 of 24 hours were counted. (Ten slides/month were counted and six traverses/slide were examined. Each traverse = 1 hour.)

assess the meaning of the fairly constant recovery of Ascomycetes spores throughout the day. A closer examination of the raw data (not shown) reveals that "Others" peak at different times of the day, similar to all the fungi recovered, depending on the month. This suggests that these fungi have monthly cycles of spore discharge in association with climate.

Cladosporium was found to be an excellent marker organism for the study summarized in Table 14–2. Although *Cladosporium* only comprises 23% of the total mould, a correlation coefficient of $r = 0.972$ was established between this mould and the total mould recovered throughout the year. Although this phenomenon is not well understood, this close relationship has also been reported in Sweden (63,72), Israel (7), and the United States (21).

In Table 14–2 the times of recovery of *Cladosporium* and *Alternaria* spores using the Burkard sampler are shown. Two peaks are seen for *Cladosporium*, one at midnight (2400 hours) and one at 1600 hours, with its minimum at 1200 hours. These data differ significantly from those presented by Gregory (35) in England, who reported maximum spore recovery at noon when using a Hirst spore trap. This difference might be due to genetic differences in spore-release mechanisms for this mould at the two locations, to climatic differences, or to both. As indicated by Ingold (48), a periodicity might be recorded that is quite unconnected with periodic spore liberation caused by differences in weather. Although Gregory's results are due to what we might term "dry spore phenomenon," this same phenomenon does not appear to be as readily apparent from the data from Tucson. In fact, the data presented is a compilation from 12 months, whereby each month presents a different picture of *Cladosporium* frequency on a daily basis (data not shown).

The *Alternaria* curve is more easily understood. There is an apparent relationship between relative humidity and temperature and sporulation

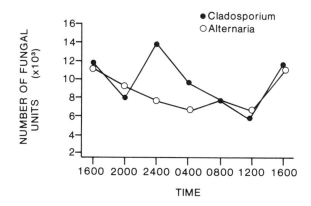

Fig. 14–3. The number of *Cladosporium* and *Alternaria* units enumerated by a Burkard sampler throughout the day for the year 1979 in Tucson.

of *Alternaria*. Over the year peak recoveries of *Alternaria* spores are associated with increasing temperature and decreasing relative humidity. This pattern can be seen on a daily basis (Fig. 14–3). During the nighttime hours, as relative humidity increases and temperature decreases, sporulation decreases, and as the morning begins the relative humidity decreases and temperature increases, leading to greater spore production.

Because it is a dry spore mould, *Cladosporium* might be expected to follow the same sporulation pattern as *Alternaria* in terms of seasonal (72) and daily relative humidity and temperature relationships. If so, what accounts for the rapid changes in *Cladosporium* counts over a day's time (24 hours) as compared to the smooth curve exhibited for *Alternaria*? There are probably two major factors that account for this, the sporulating mechanism and the spore itself. The following is hypothesized regarding *Cladosporium*:

1. Spore production and dispersal (and their opposites) may occur on an extremely rapid basis, but this may just be local, because similar results were not obtained by Gregory (35) who reported smooth recovery curves for this fungus.
2. Spores may accumulate in the atmosphere and settle in one mass, but this does not account for rapid changes from many conidia to few.
3. Spore clumping, known to occur with *Cladosporium* and not with *Alternaria*, may occur under certain atmospheric conditions.

Because these spores were counted volumetrically, absolute numbers of conidia were counted and not relative numbers of colonies, which can yield opposite results. The spores recovered at midnight and for the most part dispersed hours earlier may have undergone a rapid decline in numbers because of atmospheric conditions that prevented clumping and accumulation, such as relative lack of humidity and upper air turbulence. This would tend to keep spores aloft more and more, with a gradual decline in spore counts recorded at ground level.

At noon a reversal occurs. Spores produced and carried aloft hours earlier have accumulated in the atmosphere, and climate conditions occur such that there is clumping and rapid settling to earth of larger conidial masses. Increasing numbers of settling masses become evident until 1600 hours, when another reversal occurs. This reversal is also associated with greater conidial production than occurred earlier, because dry air spores tend to become prevalent during daytime hours.

AGRICULTURE, MOULD, AND CLIMATE

The presence of fungi as saprobes and/or parasites on vegetation has been well documented (34,35,73,99) and the dispersal of these fungi from their specific ecologic habitat is affected by numerous climatologic factors. Results of previous studies using aerometric devices (40,77,85) have shown certain associations of minimum daily temperature, prevailing winds, and total precipitation with spore recovery.

The correlation of allergenically significant fungal genera with crop growing, harvest, or postharvest seasons may allow more efficient prediction and management of fungal aeroallergen presence. The same climatic conditions that permit planting, maturation, and harvesting of crops also permit growth and dispersal of their associated fungi by the processes of agitation and aeration.

For example, in the grain belt area of the United States,

> the most common moulds are *Alternaria* and *Cladosporium (Hormoden-drum)*. The spores of these moulds begin to appear in the atmosphere in May, and rise to a peak either in July, August, September, or October. This is related to the life cycle of local crops and foliage, and to whatever fungal plant disease may be present. In most areas of the northern part of the United States, plant debris begins to increase in the late spring, with blooming trees, shrubs, and other plants. This increase in dead plant particles plus the seasonal increase in warmth and humidity leads to increased mould growth (3).

Table 14–3 lists a few crops and their associated pathogenic fungi. Most of these fungi are found in significant amounts in the atmosphere in many locations throughout the world (60, 69, 90, 97), and their infection of crops indicates their versatility as pathogens as well as saprobes. All cycles of a crop-growing season yield fungi that are potentially airborne.

Fungi do not originate in the atmosphere but are carried aloft by wind

TABLE 14–3.   Some Allergenic Fungi and Crops They Are Known to Affect

Mould	Plant
Alternaria	Tomato, potato
Aspergillus	Stored seeds
Botrytis	Bean, lettuce, tomato, onion, strawberry
Cladosporium	Cucumber, tomato, peach, and other leaf mould
Epicoccum	Fruits
Fusarium	Tomato, cotton, banana, watermelon, muskmelon, celery, cabbage, pea
Helminthosporium	Oats, corn, grasses, pear, sugar cane
Penicillium	Citrus (rot of fleshy organs)
Phoma	Beet, crucifer
Rhizopus (most important during storage, transfer, and marketing)	Strawberry, sweet potato, peach, cherry, peanut, corn
Smut	Grasses, grains

currents. Fungi that have rhizosphere associations, such as *Penicillium*, mycelia sterilia, *Mucor, Rhizopus, Fusarium, Phoma,* and a wide variety of others (34,74), may become exposed to the air during several phases of the agricultural cycle.

Vegetated areas other than those used for agriculture may also produce potentially allergenic fungi. Climatologic conditions not only determine which spores will be released and when their release will occur, but also determine how far these spores will be carried by the wind. *Sporobolomyces*, yeasts, the Ascomycetes, and the Basidiomycetes are released during periods of high humidity, while *Cladosporium, Alternaria,* and other fungal spores are released in dry air (48). Particles released in dry air are in greater numbers during the day, while those released when the relative humidity is high tend to be more abundant at night.

MAN-MADE CLIMATE AND MOULD

Cold Mist Vaporizers

Potential hazards regarding the use of indoor humidifying and cooling devices have been documented (41,50,88,89). Both moulds and yeasts were found to colonize cold mist vaporizers (aerosol generators) (41,88) and to be released with different temporal patterns that may reflect substrate specificity within these devices.

Research into this area was precipitated by patients who had reported bouts of asthma on several occasions during the operation of cold mist vaporizers (88). They had also reported perennial allergic rhinitis. Data suggested that emissions from the vaporizer, rather than their capacity to disturb indoor air, had produced the recorded increments in microbial contamination. *Rhodotorula, Penicillium, Aspergillus, Oospora, Verticillium, Geotrichum, Sporobolomyces,* and others were found in the heavily contaminated aerosol. At times yeasts were found in excess of 12,000 to 14,000 particles per m^3 (approximately 500 ft^3) of air after turning on the device.

Old-Style Air Conditioners

Human exposure to older style air-conditioning units has resulted in hypersensitivity pneumonitis and allergic lung response, as well as allergic responses in the nose, throat, eyes, ears, and skin (50). These reactions may occur some hours after exposure. The evaporator compartment has all the essentials for fungal growth: absence of ultraviolet light, moisture, and hydrocarbon food sources (petrochemical contaminants in the air). Essentially, growth is caused by moisture condensation on the evaporator coils; mould spores that come into contact with the wet coils propagate when the unit is turned off, only to be blown into the controlled environment when it is turned on. This causes the characteristic "dirty laundry" odor so common in automobile and other air conditioners.

Evaporative Coolers vs. Air Conditioners

The evaporative cooler (EC) is an energy-efficient device that is used in two thirds of homes in southwestern United States during the summer months of April or May through September or October. Because patients as well as their physicians frequently relate flare-ups of allergic respiratory symptoms to the use of ECs, a multifaceted study was undertaken in 1980 to sample homes and coolers for agents that might be significant in precipitating these symptoms. It was found that ECs provide a niche for the establishment of fungi that may act as indoor allergens. Thus, during a season when the total atmospheric mould is often reduced because of the hot dry summer, available substrate in the EC (cellulose filter pads of Aspen fibers) and the circulating pool of water within the cooler provide a microclimate for the breeding of mould.

In the investigation cited above 41 homes with either EC or central air conditioning (AC) were sampled. Petri plates were placed inside all homes for 45-minute periods, and colonies that appeared within the next 3 weeks were catalogued.

We found that the incidence of *Alternaria* spp., *Aspergillus niger*, and *Bispora* spp. was found to be increased in homes relative to outdoors by two- to fivefold after coolers were used. Although this was an average, the ratio was much higher than this for specific fungi in some homes. The number of mycelial fragments and smut cells inside homes followed trends in atmospheric incidence.

In addition, the total numbers of fungi and the numbers of fungal genera in evaporative-cooled homes exceeded the numbers in air-conditioned homes. In virtually all cases, fungi recovered from the water were the same fungi recovered from the cooler pads.

Solomon (88,89) has reported that humidifying devices may serve as dispersion sources in addition to their permissive role in facilitating fungal growth. He also noted that dwellings that afforded copious recoveries of fungi almost invariably contained free-standing and/or duct-installed humidifiers. On initial comparison, mean fungus levels in the 92 homes of subjects receiving outpatient care were notably higher than in those of ostensibly well persons (88,89). (Fungus recoveries differed at the 90% level of significance.)

The findings of Solomon were supported by our data. Mould counts were compared against the daily diaries of asthmatics living within EC homes (56). Computer-generated data revealed that asthma cluster families (families with at least one asthmatic within the EC home) with higher fungal counts indoor and/or outdoor had somewhat more eye irritation, rhinitis, and chest tightness as compared to "low-mould" families. If further research demonstrates the EC to be unsafe, a moral issue regarding the safety of these devices would be raised.

REGIONAL CLIMATES AND MOULD

Inland Vs. Coastal Areas

In San Diego (37) it was found that those fungi that occur in increasing frequency when proceeding inland from the beach to the mountains were *Cladosporium*, sterile mycelia, and *Stemphylium*, and those fungi that occur in diminishing frequency when proceeding inland are *Penicillium*, *Chaetomium*, and *Plenozythia*. Substrate in this case is probably not crop-related; rather, changing climate affects the condition of the soil, the vegetation that grows in it, the fungi associated with the vegetation, and the dissemination of fungal particles into the atmosphere (Fig. 14–1).

In another study species of the same fungal genus were found to differ in their ecological habit (86), a climate-related phenomenon. These results were similar to the findings of Al-Doory et al. in Iraq (2).

Alternaria

Sorenson et al. (90) conducted a 12-month survey of atmospheric fungal spores in five climatologically diverse sites throughout the United States and Puerto Rico. These sites included Tacoma, Miami, San Juan, Pittsburgh, and San Antonio. Findings from Tucson will be presented in this section.

Cladosporium was the predominant organism at all sites sampled by Sorenson and co-workers, occurring on an average of 64% of the Petri plates. *Aureobasidium* was the second most common fungus at four of the five sites.

Morrow et al. (60) have indicated that *Alternaria* is one of the most frequently occurring fungi at most of the sites included in their report, although it occurred relatively infrequently in Seattle and Portland. Reyes et al. (70) reported that *Alternaria* was isolated infrequently in the Philippines. Hyde et al. (47) noted that *Alternaria* is much less prevalent in Great Britain than in the central United States but is similar in frequency at coastal regions. Our data from Tucson (83) and Salinas (84) indicate that *Alternaria* is similar to *Cladosporium* in frequency in Tucson and the fifth most common mould in Salinas.

An exception to the greater frequency of *Alternaria* inland was presented by Barkai-Golan et al. (7). In Israel a 2 year volumetric survey of airborne fungi conducted in the arid inland town of Arad revealed an approximately fourfold lower spore concentration than in the coastal region of the country. Throughout the entire survey the number of fungi collected was markedly lower in Arad than in the coastal region. The three most common fungal genera at both sites were *Cladosporium*, *Al-*

ternaria, and *Penicillium,* which in each case were found in less frequency than at the inland site.

The preponderance of evidence, however, suggests that *Alternaria* is not found in abundance in coastal regions and is more common inland. This phenomenon is probably related to a combination of climatic factors.

It has been shown that on a daily basis the formation of *Alternaria* conidia occurs in the dark hours when the relative humidity is high, and their release occurs through photomechanisms when the warmth of the day begins (54,55). Thus the decrease of relative humidity and the increase of temperature lead to spore discharge.

Leach (54) has reported that spore release in *Alternaria tenuis* was triggered by lowering relative humidity from near saturation, by increasing the relative humidity from lower to higher levels, and by exposing sporulating specimens to light. Infrared, shorter red, and near

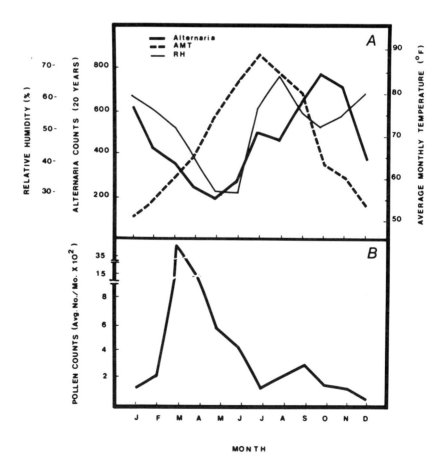

Fig. 14–4. *A.* Seasonal incidence of *Alternaria* compared with climatic variables in Tucson over a 20-year period. *B.* Seasonal incidence of pollen in Tucson.

ultraviolet wavelengths stimulated spore release in this mould (53–55). The changes in *Alternaria* counts in the months of May and October (Fig. 14–4) (83) may also be explained on the basis of the average monthly temperature, which has risen or fallen, respectively, to approximately 70° F. This temerature conforms to the optimal temperature described for differentiation of conidiophores and conidia of *Alternaria solani* (44). *A. alternata* and *A. solani* have been identified in our laboratory from Petri plates exposed to the atmosphere in Tucson. This temperature is also within the range of optimum growth for most fungi (68–75° F) (20). These recoveries link *in vivo* climatologic variables affecting spore maturation and/or release by species of *Alternaria* with experimental findings for these same moulds.

Numerous climate-mould relationships have been found to exist in arid and nonarid cities (7,12,23,24,67,77,82,90), but mould will not grow without a substrate to grow on. The increasing number of pollen-producing shade trees such as mulberry, cottonwood, and eucalyptus in Tucson over the years has not only resulted in a pollen plague every spring but can also be linked indirectly to increasing amounts of substrate supporting growth of allergenic moulds, a situation that may also be occurring in arid areas such as Kuwait (100).

Considering all the data presented, it is reasonable to assume that inland environments must provide *Alternaria* with the right combination of sunlight, temperature, humidity, and other factors that permit its greater proliferation as compared to coastal areas. Therefore, it might be contrary to expectation that there are sufferers from perennial respiratory allergy who have been found to maintain good health inland but who develop symptoms when living at or visiting the coast (64–66). This "climate group" of sufferers is found in South Africa, Israel, Spain, and warmer countries (67). In the countries studied, the climate at the coast is characterized by high relative humidity and temperature, both within a narrow range during the day and throughout the year; this contrasts with the climate pattern inland, where there is a correspondingly wide range of relative humidity and temperature.

As early as 1932 Jimenez-Diaz et al. (49) claimed that atmospheric fungi in the humid coastal air and ground are the agents responsible for the occurrence of symptoms in this "climate group" of patients. It has been Ordman's contention (64–66) that inland house dust from the interior Highveld regions of South Africa (and perhaps in other countries as well), with clear skies and long hours of bright sunshine, is acted on by ultraviolet solar radiation to reduce its allergic potential. This may not be the case in coastal regions, so a higher rate of respiratory allergy is found in the more humid areas.

Temperate Climates and Mould

From Sweden Nilsby (63) has reported that the amount of *Hormodendrum* recovered follows seasonal patterns and closely parallels the amount of total mould. Nilsby, however, could not explain the seasonal variations of this mould or of *Alternaria*, because there was no correlation with wind direction, air temperature, or the overcast. It was found, however, that there was an increase in spore sedimentation on the plates during a strong wind, especially during foggy weather.

Results of studies of aircraft flights at varying altitudes reveal that the mould spore content decreases with increasing heights up to 1000 m (approximately 3333 ft), with surprisingly high concentrations of spores in clouds (63). It is postulated that the mould spores are carried aloft with the rising air currents and, when the moisture in these currents condenses, the mould spores adhere to the fine water particles, a situation similar to that believed to occur during a fog. The relationships between high altitude and microorganisms has been presented in a number of other papers in more detail (30–33,38).

Ripe (72) reported that *Cladosporium* was the predominant fungus at all the outdoor exposure points in Stockholm, and closely paralleled the total mould count as reported by Nilsby (63). This mould was reported to follow weather changes closely, both indoors and outdoors, similar to findings elsewhere. It appears from Ripe's data (72) that, as the temperature increases and spring approaches, the relative humidity decreases and the mould counts go up, and as the temperature decreases later in the year and the relative humidity increases the mould counts go down.

A mould season quite similar to that of Sweden has been demonstrated in Denmark (29).

In Delhi mould allergy does not appear to be a significant problem, because skin tests using 29 fungal antigens on almost 1097 patients revealed only an average of 5.9% of patients with 2+ to 4+ reactions (1,80). The percentage of Delhi patients showing symptoms of nasobronchial allergy, however, indicates a highly significant negative correlation ($r = -0.96$) with the mean monthly temperature (6).

In Greece, Bartzokas (8) concluded that an increase in temperature results in a reduction of the population of airborne fungi. When the relative humidity and the rainfall increase, so does the concentration of air flora. The variability of the air fungi population, at a level of 75.3%, depends on the variations in the values of the meteorologic conditions, the most important of which are humidity (positive correlation) and temperature (negative correlation).

In Johannesburg it has been reported (67) that for *Cladosporium* and *Alternaria* there is no obvious seasonal incidence. This may be related to the fact that throughout the year the morning mean air temperature only varies by 23° F, while the afternoon air temperature varies by 11° F. It is therefore possible that fairly constant climatologic factors are responsible for the nonseasonal incidence of the fungi reported. At an altitude of 5600 feet, Johannesburg is also essentially unaffected by the flora at lower altitudes. In addition, a noted lack of mould substrate in the form of agricultural commodities may help contribute to the lack of seasonality of seasonal fungi.

Tropical Areas

Mould spores are especially important in the tropics, in which climatic conditions are very favorable to the growth of fungi and may result in a high concentration of spores in the air, which in turn causes an increased incidence of allergic diseases (62).

For regions between sea level and 1000 feet, annual rainfall reaches 2000 to 3000 mm (80–120 inches), with a median annual atmospheric humidity over 90% and a median annual temperature of approximately 76° F.

In all countries studied (62) *Penicillium* and *Aspergillus* were recovered. *Hormodendrum* is found frequently in low and humid cities such as Havana (10,11) and Rio de Janeiro (62), and in Panama (94). The Phycomycetes, such as *Rhizopus* and *Mucor,* are more frequent in higher locales such as Quito, Ecuador. *Alternaria* is more common in noncoastal areas (90).

In Panama there is a general correlation between the curves representing the total mould counts and the mean relative humidity, and not with mean monthly temperature. It is the opinion of Taylor and McFadden (94) that the relatively uniform annual temperature in Panama (80–85° F) minimizes this environmental factor as a major contributor to the varying airborne mould flora.

A similar situation has been presented in a report from Hong Kong (97), but in this instance it is the relative humidity that remains reasonably constant throughout the year (75–80%), and it is the mean daily temperature that is the variable. The counts of fungal air flora, in general, fluctuate around the changing temperature and increase sharply when the temperature rises.

In Havana (11) results of a study of 1200 cases of rhinitis and asthma strongly suggest that the periods of high contamination of the air with spores of common moulds coincide with periods when allergic symptoms are most severe, usually the beginning of the rainy season.

It must be stressed that at least one climatic variable must fluctuate to produce seasonality of fungal flora, given constancy of substrate, or the substrate must vary (as with plant pathogens) if one or more climatic variables are to remain constant to cause regular fluctuations in airborne fungal flora.

Arid Climates and Mould

Davies (20) has reported that in Kuwait, between August and November, as the monthly accumulated temperature below 29° C rises, the concentration of *Cladosporium* (66% of the total mould) increases to its seasonal high and reaches a maximum in November. From November onward the number of days below 29° C increases, and the concentration of *Cladosporium* then decreases.

Davies (20) has also reported that *Cladosporium* conidia are detached from conidiophores by the impact of minute water droplets, resulting in the higher incidence of atmospheric fungi in more humid coastal regions. In arid climates the effect of windblown sand will be similar to that of water droplets (20), and probably accounts for the longer lengths of hyphae, conidiophores, and other fungal elements on collection devices.

When there is cultivation of an area and extension of vegetative cover, the appearance of allergy among the inhabitants becomes more commonplace. This is also true in developing arid communities. In fact, vegetation, population, pollen, mould, and allergy to pollen and mould may increase in a parallel manner (68,83,87).

In another study from Kuwait, Moustafa and Kamel (61) reported that *Alternaria* counts were highest when the mean temperature fell between 20 and 34° C and relative humidity was between 28 and 45%.

The composition of the fungal air flora of Kuwait is determined mainly by prevailing winds that carry fungal spores from Iraq to the northwest and Iran to the north and southwest, both agriculturally oriented countries that grow cereals and grasses. The four most common moulds recovered from the air of Iraq were *Aspergillus, Penicillium, Alternaria,* and *Cladosporium* (4), also the most commonly identified moulds in the air of Kuwait.

The high incidence of *Aspergillus* spores reported in the atmosphere in Iran (5,36,79) and in Iraq helps to explain the fact that immunodiffusion tests have shown a number of serologically positive patients for *Aspergillus* infection in the Tehran area.

The number of fungi cultured from the air of Arad, Israel, an arid community, was only 27% of the number recovered from Rehovat, a coastal city in Israel (7).

Figure 14–5 shows some of these Middle Eastern countries and their geographic relation to one another.

In spite of all the evidence relating mould prevalence to changes in temperature and relative humidity, lack of substrate is most likely the limiting factor for fungal growth. If agriculture and farmland develop, it is also probable that the low humidity will not significantly retard the development and propagation of large amounts of allergenic fungi, and that subsequently the indigenous asthmatic population will increase.

MOULD ALLERGY AND BIOCLIMATOLOGY

The literature pertaining to the direct effect of climate on the increased susceptibility of individuals to various respiratory disorders is extensive.

Fig. 14–5.   Some Middle Eastern countries that have reported climate-mould relationships.

The research of Tromp (95,96) has gone far toward consolidating information scattered for long periods of time.

The specific relationships between climate sensitization and mould allergy have been alluded to in the literature with much less frequency. To emphasize the complexity of the issue, note that although mould allergy involves the IgE immune response and may precipitate an asthmatic attack, asthma is strictly respiratory, and an attack is not necessarily precipitated by mould and need not be IgE-mediated. A cause-effect relationship between weather conditions that enhance mould allergy in an individual may or may not provoke asthma in that same person.

The four climatic factors that influence the development and severity of allergic conditions are barometric pressure, humidity, altitude topography, and temperature (92). Also, if the specific reference is to mould allergy, then the amount and type of mould must be considered, as well as the presence or absence of wind. In addition, climatic factors, such as the combination of rapidly changing barometric pressure and humidity, may act synergistically to make someone more sensitive to mould than any one factor alone would (22,42,81).

Criep and Hammond (16) have found that

> The allergic response is related largely to the function of the capillary bed, especially in manifestations of superficial structures such as the nasal and bronchial mucous membranes and the skin. It is easy to see then that changes in barometric pressure will exercise some influence on the calibre of the capillaries, and hence on capillary permeability. Such changes in barometric pressure are associated with stormy weather and aggravate allergic conditions.

What Criep and Hammond are probably observing is an interaction of climatic factors and human susceptibility to allergic substances, because storm fronts can be associated not just with changes in barometric pressure but with *rapidly changing* barometric pressure and relative humidity, electrical discharges in the atmosphere, winds bearing high concentrations of spores, dust, and positive ions, cold snaps, precipitation, optical associations, sunlight and darkness affecting behavior and attitude, and other factors. Storms also cause persons to move indoors, where heating ducts are either blowing out dust and mould spores or are producing wind currents that stir pre-existing dust. The absolute temperature difference between the outdoor and indoor environment may also be significant enough to precipitate an allergic response.

Salvaggio et al. (75) have reported that high "asthma admission" days at New Orleans Charity Hospital were associated with minimum temperature and minimum relative humidity during the months of September through November, as well as with low wind speed and southwest wind direction (October only).

Heise and Heise (39) have shown that temperature changes, with altitude, may be factors in the rise and fall of the number of pollens and moulds in the atmosphere. Although high altitude has been recommended for many years as a possible source of relief for the asthmatic, many persons at high altitude suffer from various allergies because of spores being carried aloft. This is similar to an effect noted by Davies (20). Air temperatures in city streets are usually warmer than those in higher air masses, which causes the uplift of spores. In fact, the patient near an open window on the top floor of a building that rises above the general level of the rooftops in its vicinity is exposed to higher pollen and spore concentrations than those that are found at street level.

Hot winds (above body temperature) dry the mucous membranes of the respiratory tract, thus making the body more susceptible to allergenic substances. These winds also create an excess of positive ions, reported to be a deleterious effect when interacting with certain cells in the body (52,92,93,95,96).

In 1917 Scheppegrell (78) observed that hay fever patients improve during a rain of several days' duration, and that their symptoms become aggravated when high winds prevail.

Increased humidity and fog influence respiratory allergy adversely, probably because of the increased breathing resistance produced by the moisture particles in the air. Furthermore, the conditions leading to the accumulation of fog also produce high concentrations of particulate contaminants (16,20,63).

In conclusion, considering the array of data available, climatotherapy should be a highly individualized form of treatment that is of unquestionable value when patients are carefully selected. The wholesale shipment of allergics and asthmatics to developing cities over a period of decades by some physicians has led to populations of genetically susceptible senior citizens and their children, who suffer greatly when the pollen and mould season arrives.

REFERENCES

1. Agarwal, M.K., Shivpuri, D.M., and Mukerji, K.G.: Studies on the allergenic fungal spores of the Delhi, India, metropolitan area. J. Allergy, **44**:193, 1969.
2. Al-Doory, Y., Tolba, M.K., and Al-Ani, H.: On the fungal flora of Iraqi soils. II. Central Iraq. Mycologia, **51**:429, 1959.
3. Allergy Foundation of America: Allergy to Molds. New York, Allergy Found. America.
4. Al-Tikriti, S.K., Al-Salihi, M., and Gaillard, G.E.: Pollen and mold survey of Baghdad, Iraq. Ann. Allergy, **45**:97, 1980.
5. Amin, R., and Bokhari, M.H.: Survey on atmospheric fungus spores in Shiraz, Iran (1977). Ann. Allergy, **42**:246, 1979.
6. Babu, C.R., Singh, A.B., and Shivpuri, D.M.: Allergenic factors and symptomatology of respiratory allergy patients. J. Asthma Res., **16**:97, 1979.
7. Barkai-Golan, R., et al.: Atmospheric fungi in the desert town of Arad and in the coastal plain of Israel. Ann. Allergy, **38**:270, 1977.
8. Bartzokas, C.A.: Relationship between the meterorological conditions and the airborne fungal flora of the Athens metropolitan area. Mycopathol., **57**:35, 1975.
9. Bernton, H.S.: Asthma due to a mold—*Aspergillus fumigatus*. JAMA, **95**:189, 1930.
10. Cadrecha-Alvarez, J., and Fernandez, J.: Quantitative studies of air-borne fungi of Havana in each of the twenty-four hours of the day. J. Allergy, **23**:259, 1952.
11. Cadrecha-Alvarez, J., and Fernandez, J.; Numbers and kinds of air-borne, culturable fungus spores in Havana, Cuba. J. Allergy, **26**:150, 1955.
12. Cammack, R.H.: Seasonal changes in three common constituents of the air spora of Southern Nigeria. Nature, **176**:1270, 1955.
13. Chatterjee, J., and Hargreave, E.: Atmospheric pollen and fungal spores in Hamilton in 1972 estimated by the Hirst automatic volumetric spore trap. Can. Med. Assoc. J., **110**:659, 1974.
14. Collins-Williams, C., et al.: Atmospheric mold counts in Toronto, Canada, 1971. Ann. Allergy, **31**:69, 1973.
15. Credille, B.A.: Report of a case of bronchial asthma due to molds. J. Mich. Med. Soc., **32**:167, 1933.
16. Criep, L.H., and Hammond, M.L.: Regional factors in allergy. Ann. Allergy, **10**:282, 1952.
17. Davies, R.R.: Detachment of conidia by cloud droplets. Nature, **183**:1695, 1959.
18. Davies, R.R.: Viable moulds in house dust. Trans. Br. Mycol. Soc., **43**:617, 1960.
19. Davies, R.R.: Climate and topography in relation to aeroallergens at Davos and London. Acta Allergy (Copenh.), **24**:396, 1969.
20. Davies, R.R.: Spore concentrations in the atmosphere at Ahmadi, a new town in Kuwait. J. Gen. Microbiol., **55**:425, 1969.
21. Deamer, W.C., and Graham, H.W.: Respiratory mold allergy—a 12-month atmospheric survey in San Francisco, Calif. West. J. Med., **66**:289, 1947.
22. Dingle, A.M.: Meterorological considerations in ragweed hay fever research. Fed. Proc., **16**:615, 1957.
23. Dransfield, M.: The fungal air-spora of Saman, Northern Nigeria. Trans. Br. Mycol. Soc., **49**:121, 1966.
24. Dupont, E.M., et al.: A survey of the airborne fungi in the Albuquerque, New Mexico, metropolitan area. J. Allergy Clin. Immunol., **39**:238, 1967.
25. Durham, O.C.: Atmospheric allergens in Alaska. J. Allergy, **12**:307, 1941.
26. Dworin, M.: A study of atmospheric mold spores in Tucson, Arizona. Ann. Allergy, **24**:31, 1966.
27. Fein, B.T., and Kamin, P.B.: The climatic factors in the etiology of allergic diseases. J. Asthma Res., **3**:17, 1965.
28. Feinberg, S.M.: Mold allergy: Its importance in asthma and hay fever. Wis. Med. J., **34**:254, 1935.

29. Flensborg, E.W., and Samsoe-Jensen, T.: Studies in mold allergy. 3. Mold spore counts in Copenhagen. Acta Allergy (Copenh.), **3**:39, 1950.
30. Fulton, J.D.: Microorganisms of the upper atmosphere. III. Relatioinship between altitude and micropopulation. Appl. Microbiol., **14**:237, 1966.
31. Fulton, J.D.: Microorganisms of the upper atmosphere. IV. Microorganisms of a land air mass as it traverses an ocean. Appl. Microbiol., **14**:241, 1966.
32. Fulton, J.D.: Microorganisms of the upper atmosphere. V. Relationship between frontal activity and the micropopulation at altitude. Appl. Microbiol., **14**:245, 1966.
33. Fulton, J.D., and Mitchell, R.B.: Microorganisms of the upper atmosphere. II. Microorganisms in two types of air masses at 690 meters over a city. Appl. Microbiol., **14**:232, 1966.
34. Geary, W.D.: The Relation of Fungi to Home Affairs. New York, Henry Holt and Co., 1959.
35. Gregory, P.H.: Microbiology of the Atmosphere. 2nd Ed. New York, John Wiley and Sons, 1973.
36. Hariri, A.R., et al.: Airborne fungal spores in Ahwaz, Iran. Ann. Allergy, **40**:349, 1978.
37. Harsh, G.F., and Allen, S.E.: A study of the fungus contaminants of the air of San Diego and vicinity. J. Allergy, **16**:125, 1945.
38. Heise, H.A., and Heise, E.R.: The influence of temperature variations and winds aloft on the distribution of pollens and molds in the upper atmosphere. J. Allergy, **20**:378, 1949.
39. Heise, H.A., and Heise, E.R.: Effect of a city on the fallout of pollens and molds. JAMA, **163**:803, 1957.
40. Hirst, J.M.: Changes in atmospheric spore content: Diurnal periodicity and the effects of weather. Trans. Br. Mycol. Soc., **36**:375, 1953.
41. Hodges, G.R., Fink, J.N., and Schlueter, D.P.: Hypersensitivity pneumonitis caused by a contaminated cool-mist vaporizer. Ann. Intern. Med., **80**:501, 1974.
42. Hollander, J.E.: Environment and musculoskeletal diseases. Arch. Environ. Hlth., **6**:527, 1963.
43. Hopkins, J.G., Benham, R.W., and Kesten, B.M.: Asthma due to a fungus—*Alternaria*. JAMA, **94**:6, 1930.
44. Horsfall, J.G., and Lukens, R.J.: Differential temperatures for separate phases of *Alternaria solani*. Phytopathol., **61**:129, 1971.
45. Hudson, H.J.: Fungal Saprophytism. London, Edward Arnold, 1972.
46. Hyde, H.A., Richards, M., and Williams, D.A.: Allergy to mould spores in Britain. Br. Med. J., **1**:886, 1956.
47. Hyde, H.A., and Williams, D.A.: The incidence of *Cladosporium herbarum* in the outdoor air of Cardiff. Trans. Br. Mycol. Soc., **36**:260, 1953.
48. Ingold, C.T.: Spore Liberation. Belfast. London, Oxford University Press, 1965.
49. Jimenez-Diaz, C.C., Sanchez Cuenca, B., and Puig, J.: Climate asthmas. J. Allergy, **3**:396, 1932.
50. Kelsey, P.: A look at fungus growth on a/c units: Its cause, effects, possible remedies. Air Conditioning, Heating and Refrigeration News, October 21, 1974.
51. Kethley, T.W., and Cown, W.B.: In defense of settling plates. Hosp. Manage., **103**:84, 1967.
52. Krueger, A.P., and Reed, E.J.: Biological impact of small air ions. Science, **193**:1209, 1976.
53. Langenberg, W.J., Sutton, J.C., and Gillespie, T.J.: Relation of weather variables and periodicities of airborne spores of *Alternaria dauci*. Phytopath., **67**:879, 1977.
54. Leach, C.M.: Influence of relative humidity and red-infra-red radiation on violent spore release by *Drechslera turcica* and other fungi. Phytopath., **65**:1303, 1975.
55. Leach, C.M.: Sporulation of diverse species of fungi under near ultraviolet radiation. Can. J. Bot., **40**:151, 1975.
56. Lebowitz, M.D., et al.: The adverse effects of biological aerosols, other aerosols, and micro-climate indoors on asthmatics and nonasthmatics. Presented at the Intern. Symp. on Indoor Air Pollution, Health and Energy Conservation, Amherst, MA, October 14, 1981.
57. McDevitt, T.J., et al.: Allergic evaluation of cereal smuts. Ann. Allergy, **38**:12, 1977.

58. Moore-Landecker, E.: Fundamentals of the Fungi. Englewood Cliffs, NJ, Prentice-Hall, 1972.
59. Morrow, M.B., Lowe, E.P., and Prince, H.E.: Mold fungi in the etiology of respiratory allergic diseases. I. Survey of airborne molds. J. Allergy, **13**:215, 1941.
60. Morrow, M.B., Meyer, G.H., and Prince, H.E.: Summary of air-borne mold surveys. Ann. Allergy, **22**:575, 1964.
61. Moustafa, A.F., and Kamel, M.: A study of fungal spore populations in the atmosphere in Kuwait. Mycopathol., **59**:29, 1976.
62. Naranjo, P.: Etiological agents of respiratory allergy in tropical countries of Central and South America. J. Allergy, **29**:362, 1958.
63. Nilsby, I.: Allergy to moulds in Sweden: A botanical and clinical study. Acta Allergy (Copenh.), **2**:57, 1949.
64. Ordman, D.: Respiratory allergy in the coastal areas of South Africa: The significance of climate. S. Afr. Med. J., **29**:173: 1955.
65. Ordman, D.: Relation of climate to respiratory allergy. Ann. Allergy, **19**:29, 1961.
66. Ordman, D.: Respiratory allergy and the regional climate in Israel. Ann. Allergy, **25**:106, 1967.
67. Ordman, D., and Etter, K.G.: The air-borne fungi in Johannesburg. S. Afr. Med. J., Nov., 1956.
68. O'Rourke, M.K.: Pollen dispersal and its relationship to respiratory illness. Presented at the 1st Intern. Cong. on Aerobiology, Munich, August 13–15, 1978.
69. Prince, H.E., and Meyer, G.H.: An up-to-date look at mold allergy. Ann. Allergy, **37**:18, 1976.
70. Reyes, A.C., et al.; Atmospheric mold spore count in Cebu City, Philippines. Southeast Asian J. Trop. Med. Pub. Hlth., **1**:263, 1970.
71. Richards, M.: Atmospheric mold spores in and out of doors. J. Allergy, **25**:429, 1954.
72. Ripe, E.: Mould allergy. I. An investigation of the airborne fungal spores in Stockholm, Sweden. Acta Allergy (Copenh.), **17**:130, 1962.
73. Robinson, R.K.: Ecology of Fungi. London, The English University Press, 1961.
74. Rouatt, J.W., Peterson, E.A., and Katznelson, H.: Microorganisms in the root zone in relation to temperature. Can. J. Microbiol., **9**:227, 1963.
75. Salvaggio, J., et al.: New Orleans asthma. II. Relationship of climatologic and seasonal factors to outbreaks. J. Allergy, **45**:257, 1970.
76. Salvaggio, J., and Seabury, J.: New Orleans asthma. IV. Semi-quantitative airborne spore sampling, 1967 and 1968. J. Allergy Clin. Immunol., **48**:82, 1971.
77. Salvaggio, J., Seabury, J., and Schoenhardt, E.A.: New Orleans asthma. V. Relationship between Charity Hospital asthma admission rates, semi-quantitative pollen and fungal spore counts, and total particulate aerometric sampling data. J. Allergy Clin. Immunol., **48**:96, 1971.
78. Scheppegrell, W.: Hay fever and hay fever pollens. Arch. Intern. Med., **19**:959, 1917.
79. Shafiee, A., and Rahmini, T.: Atmospheric mold spores in Tehran, Iran. Ann. Allergy, **40**:138, 1978.
80. Shivpuri, D.M., and Agarwal, M.K.: Studies on the allergenic fungal spores of the Delhi, India, metropolitan area: Clinical aspects. J. Allergy, **44**:204, 1969.
81. Sneller, M.R.: The AllergIndex—an eight-variable meteorological system designed to forecast the state of health of allergics, asthmatics, arthritics, and otherwise normal individuals. Aeroallergen Research, Tucson, 1981.
82. Sneller, M.R., and Pinnas, J.L.: Fungal ecology of the evaporative cooler (in manuscript), 1983.
83. Sneller, M.R., Hayes, H.D., and Pinnas, J.L.: Frequency of airborne *Alternaria* spores in Tucson, Arizona, over a 20-year period. Ann. Allergy, **46**:30, 1981.
84. Sneller, M.R., and Roby, R.R.: Incidence of fungal spores at the homes of allergic patients in an agricultural community. I. A 12-month study in and out of doors. Ann. Allergy, **43**:225, 1979.
85. Sneller, M.R., Roby, R.R., and Thurmond, L.M.: Incidence of fungal spores at the homes of allergic patients in an agricultural community. III. Associations with local crops. Ann. Allergy, **43**:352, 1979.
86. Sneller, M.R., and Swatek, F.W.: Distribution of the genus *Cryptococcus* in Southern California soils. Sabouraudia, **12**:46, 1974.

87. Solomon, A.M., and Hayes, H.D.: Impacts of urban development upon allergic pollen in a desert city. J. Arid Environ., **3**:169, 1980.
88. Solomon, W.R.: Fungus aerosols arising from cold-mist vaporizers. J. Allergy Clin. Immunol., **54**:222, 1974.
89. Solomon, W.R.: Assessing fungus prevalence in domestic interiors. J. Allergy Clin. Immunol., **56**:235, 1975.
90. Sorenson, W.G., Bulmer, G.S., and Criep, L.H.: Airborne fungi from five sites in the continental United States and Puerto Rico. Ann. Allergy, **33**:131, 1974.
91. Staley, J.T., Palmer, F., and Adams, J.B.: Microcolonial fungi: Common inhabitants on desert rocks? Science, **215**:1093, 1982.
92. Sulman, F.G.: Health, weather and climate. *In* Perspectives in Medicine. Vol. 7. Edited by S. Basel, A.G. Karger, 1976.
93. Sulman, F.G., et al.: Air-ionometry of hot, dry desert winds (Sharav) and treatment with air ions of weather-sensitive subjects. Int. J. Biometeorol., **18**:313, 1974.
94. Taylor, R.L., and McFadden, A.W.: Survey of airborne mold flora in Panama. Mycopathol. Mycol. Appl., **17**:159, 1962.
95. Tromp, S.W.: Influence of weather and climate on asthma and bronchitis. Rev. Allergy, **22**:1027, 1968.
96. Tromp, S.W.: Influence of weather and climate on allergy. *In* Progress in Biometeorology. Edited by S.W. Tromp. Amsterdam, Swets and Zeitlinger, 1977.
97. Turner, P.D.: The fungal air-spora of Hong Kong as determined by the agar plate method. Trans. Br. Mycol. Soc., **49**:255, 1966.
98. Waldbott, G.L., and Asher, M.S.: Rust and smut, major causes of respiratory allergy. Ann. Intern. Med., **14**:215, 1940.
99. Wheeler, B.E.J.: Fungal parasites of plants. *In* The Fungi. Vol. 3—An Advanced Treatise. Edited by G.C. Ainsworth and A.S. Sussman. New York, Academic Press, 1968.
100. Wilkinson, W.M.: Development of allergy in the desert. J. Trop. Med. Hyg., **67**:16, 1964.

MYCOLOGIC TERMS

Actogenous: Borne at the tip. Applied to spores that develop at the tip of a conidiophore.

Aerial hyphae: Hyphae above the agar surface on which the fungus is grown.

Aerobic: Requiring the presence of oxygen for survival.

Aleuriospore: Conidium.

Amerospore: Single-celled asexual spore.

Anaerobic: Requiring no oxygen for survival.

Antheridium: Fungal structure containing male nuclei (Phycomycetes).

Anthrophilic: Refers to an organism that infects humans.

Apothecium: Open, cup-shaped ascocarp.

Arthroconidium: Arthrospore.

Arthrospore: Asexual spore produced by fragmentation of a hypha; separation occurs at a septum.

Ascocarp: Structure that bears asci (Ascomycetes).

Ascomycetes: Class of fungi distinguished by its sexual reproduction through the formation of ascospores.

Ascospore: Sexual spore borne in an ascus (Ascomycetes).

Ascus: Saclike structure in which ascospores are developed through sexual reproduction (Ascomycetes).

Ballistoconidium: Ballistospore.

Ballistospore: Conidium that is forcibly discharged from its conidiogenous cell.

Basidiocarp: Structure that bears basidia (Basidiomycetes).

Basidiomycetes: Class of fungi distinguished by its sexual reproduction through the formation of basidiospores.

Basidiospore: Sexual spore borne on a basidium (Basidiomycetes).

Basidium: Club-shaped structure on which basidiospores are developed through sexual reproduction (Basidiomycetes).

Biphasic: Dimorphic.

Blastoconidium: Blastospore.

Blastospore: Asexual spore produced by budding from mycelia or pseudomycelia.

Budding cell: Any cell in any yeast from which buds are produced at one or more points.

Capsule: Mucopolysaccharide hyaline sheath surrounding the wall of a cell.

Carotenoid pigment: Red or yellow pigment containing one or more carotenoid compound; found usually in the yeast *Rhodotorula*. It may also present in some filamentous fungi.

Chlamydospore: Thick-walled spore characterized by a high concentration of stored materials.

Chlorophyll: Green pigment of algae and green plants through which these plants produce their own food supplies (photosynthesis).

Clavate: Club-shaped.

Cleistothecium: Closed ascocarp.

Coenocytic: Refers to aseptate hyphae.

Columella: Terminal swollen segment of a sporangiophore that pushes up into the sporangium.

Conidiogenous cell: Cell that produces conidia.

Conidiophore: Specialized hypha on which conidia are borne.

Conidium: Asexual spore borne in various ways by fungi.

Coremia: Fruiting body consisting of a sterile stalk of parallel hyphae and a terminal head of fertile or spore-bearing branches.

Cross wall: Septum; found in hyphae or spores.

Cutaneous mycosis: Mycotic disease infecting the skin, hair, or nails.

Dematiacious: Refers to a fungus having brown to black pigmentation.

Dermatomycosis: Cutaneous disease caused by any fungus.

Dermatophyte: Fungus (parasitic) that can use keratin (nails, skin, hair) as a source of food.

Dermatophytosis: Mycotic disease caused by a dermatophyte.

Dichotomous: Branching in two directions.

Dictyospore: Multicelled asexual spore divided into two or more planes by septa.

Didymospore: Two-celled asexual spore.

Dimorphic: Refers to a fungus with a yeast form (in host) and a mould form (in nature); found mostly among pathogenic fungi.

Disseminated mycosis: Mycotic disease that involves more than two sites of the body.

Downy: Covered with short, sparse-to-dense, aerial hyphae.

Encapsulated: Refers to cells that are surrounded by capsules.

Endogenous: Refers to infection originating within the body.

Endospore: Asexual spore developed within a cell.

Eumycota: True fungi.

Exogenous: Refers to infection originating outside the body.

Family: Taxonomic division of an order containing related genera.

Filament: Hypha.

Filamentous: Refers to a fungus composed of hyphae.

Floccose: Wooly or cottony.

Fungi imperfecti: Fungi that reproduce asexually only.

Fungus: Any member of the Kingdom Mycetaea.

Fusiform: Spindle-shaped.

Gametangium: Structure carrying a gamete (Phycomycetes).

Genus: Taxonomic division of a family containing related species.

Geophilic: Fungi whose natural habitat is in the soil.

Germ tube: Tubelike outgrowth from a germinating spore that develops into a hypha.

Heterotrophic: Using organic compounds as energy sources.

Hyaline: Colorless.

Hypha: Single filament of the mycelium of a fungus.

Imperfect fungus: Fungus that reproduces by asexual methods only.

Imperfect stage: Asexual stage.

Intercalary: Formed between cells of a hypha, especially those of chlamydospores.

Macroaleuriospore: Macroconidium.

Macroconidium: Larger of two types of conidia produced by the same fungus.

Mastigomycotina: Subdivision of true fungi characterized by their flagellated zoospores.

Meiosis: Cell division in which new cells have half the number of the original chromosome—reduction division (sexual reproduction).

Metulae: Branches from conidiophores that carry the sterigmata *(Penicillium)*.

Microaleuriospore: Microconidium.

Microconidium: Smaller of two types of conidia produced by the same fungus.

Mitosis: Cell division in which new cells have the same number of chromosomes (asexual reproduction).

Monomorphic: Having only one form.

Mould: Filamentous fungus.

Mucilaginous: Slimy.

Muriform: Having both horizontal and vertical cross walls.

Mycelia sterilia: Hyphae that produce neither conidia nor spores.

Mycelium: Mass of hyphae that form the thallus of fungi.

Mycosis: Disease caused by a fungus.

Myxomycetes: Group of fungi characterized by amoeboid stages in their life cycle.

Nodular body: Round ball-like structure formed from a mass of intertwined hyphae.

Oogonium: Structure that contains one or more female gametes (Phycomycetes).

Oospore: Zygote formed by union of two unlike gametes (Phycomycetes).

Order: Taxonomic division of a class containing related families.

Ostiole: Opening at the distal end of the pycnidium or perithecium.

Pectinate: Comblike.

Pedicel: Any slender stalk, especially one that supports a fruiting or spore-bearing organ.

Penicillate: Having a brushlike structure *(Penicillium)*.

Perithecium: Flask-shaped ascocarp open at the distal end by an ostiole.

Phialide: Subcylindric, narrow, or broadly flask-shaped cell that produces spores from the tip. The lip of the cell may be enlarged in some species forming a narrow to wide collarette.

Phialoconidium: Phialospore.

Phialospore: Spore that develops from a phialide.

Photosynthesis: Process in green plants and algae that produces food by combining water and carbon dioxide with the assistance of chlorophyll and sunlight.

Phycomycetes: Group of fungi characterized by their coenocytic hyphae.

Pleomorphism: Mutational change in a fungal colony that makes it sterile (lack of spore formation).

Pleurogenous: Borne on the sides of a conidiophore or hypha.

Polymorphic: Refers to a fungus with more than one morphologic form.

Pseudohyphae: Part of the pseudomycelium.

Pseudomycelium: A false mycelium formed in the yeasts by budding cells that remain together in chains.

Pycnidium: Closed flask-shaped structure in which asexual spores are produced and released through an ostiole.

Pyriform: Pear-shaped.

Racket hyphae: Chain of club-shaped cells, in which the larger end of one cell is attached to the narrow end of the other.

Rhizoid: Refers to a rootlike structure on the mycelium (Phycomycetes).

Saprobe: Organism that can obtain nourishment from dead organic matter.

Saprophyte: Saprobe.

Sclerotium: Hard, compact, somatic mass of mycelium.

Septum: Cross wall in the cells or hyphae of fungi.

Sessile: Formed directly on a hypha without a stalk.

Slime moulds: Myxomycetes.

Species: Taxonomic division of a genus that may contain related varieties.

Spiral hyphae: Hyphae with coiled or corkscrewlike portions.

Sporangiole: Small two or three-spored sporangium, borne externally on the specialized tip of a sporangiophore (Phycomycetes).

Sporangiophore: Stalk-like structure terminated by a sporangium (Phycomycetes).

Sporangiospore: Asexual spore formed within a sporangium.

Sporangium: Asexual, closed structure where sporangiospores are formed.

Spore: Reproductive body (unit) produced by fungi.

Sporodochium: Cushion-shaped aggregate of conidiophores.

Sporophore: Structure on which spores are produced.

Sterigma: Structure on a basidium on which basidiospores are produced. Also used in fungi imperfecti as a synonym for phialide.

Stroma: Cushionlike mat of fungal cells.

Subcutaneous mycosis: Mycotic disease infecting cutaneous and subcutaneous tissues.

Symbiosis: Living together of unlike organisms, to the benefit of both.

Synnematium: Stiffly cemented cluster of elongated conidiophores.

Thalloconidium: Thallospore.

Thallospore: Asexual spore derived from a vegetative cell of the thallus.

Thallus: Entire fungus organism.

Tuberculate: Covered with knoblike bumps.

Umbilicate: Having a depressed center.

Vegetative: Refers to the food-absorbing hyphal portion of a fungus.

Verrucose: Covered with wartlike structures.

Versicolor: Having several colors.

Verticillate: Branched in whorls of three or more.

Vesicle: Swollen or bladderlike cell *(Aspergillus)*

Yeast: Organism in which the primary method of reproduction is by budding.

Zoophilic: Generally refers to fungi infecting lower animals rather than humans.

Zoospore: Motile spore with flagella.

Zygospore: Zygote formed by the union of two similar gametes (Phycomycetes).

Zygote: Product of fusion of a male and a female cell or nucleus.

INDEX

Page numbers in *italics* refer to illustrations;
   page numbers followed by 't' refer to tables.

273